First Edition, First Printing

Contents

“This book masterfully weaves together quantum mechanics, cosmology, and simulation theory in a way that feels both academically rigorous and utterly accessible. It reenergized my passion for exploring the uncharted boundaries of modern physics.”

- ***Dr. Margaret Hathaway***, Theoretical Physicist

“Every chapter resonates with cutting-edge discussions on quantum computation and AI, highlighting intriguing possibilities for the future of technology. It’s refreshing and visionary, reminded me why I entered this field in the first place.”

- **Prof. Kenji Saito**, Computer scientist

“Rarely do we find a work that so effortlessly blends scientific depth with compelling storytelling. It pulled me in from page one, and I left feeling both informed and inspired to write in new, fearless ways.”

- **Alicia Domínguez,** Author of *Day by Day*

“A treasure trove for anyone enthralled by quantum field theory and the notion of a digitized cosmos. The chapters on quantum entanglement and the universe’s ‘code’ made me rethink accepted norms.”

- **Dr. Mikhail Petrov**, Quantum Theorist

“Perhaps, we do live in a simulation.”

- **Hans Jürgen Weber**, Mathematician

Cyrus King

The Quantum Illusion Of Reality

The Simulation Theory

"When you know nothing matters, the universe is yours"

Rick Sanchez

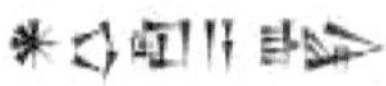

Introduction

Contradiction is a fundamental element of existence.

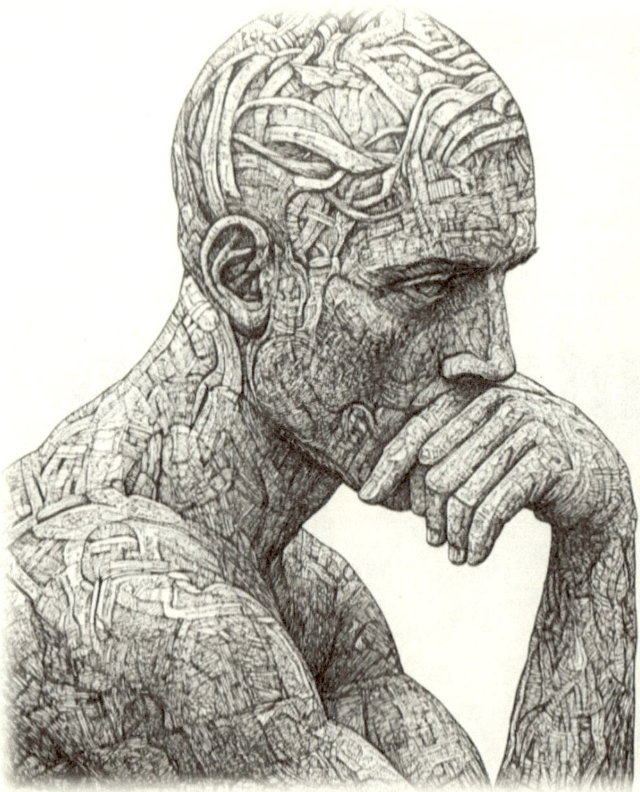

Sowing

For as long as I can remember, the night sky has inspired a profound sense of wonder. Lying beneath its endless expanse, I would often find myself consumed by a singular, overwhelming question: What is the true nature of our existence? Is the universe, as many religions suggest, an intricate design crafted by a divine hand? Or is it the breath-taking result of natural phenomena, as science strives to explain? Despite centuries of philosophical and scientific inquiry, the origin of reality remains a puzzle, a tapestry of incomplete pieces, forever tantalizing yet elusive.

Consider humanity thousands of years ago, in an era untouched by modern conveniences: no movies, no YouTube, no Google Maps or Amazon. Stories of creation were not just accepted; they made sense in the context of the time. These tales aligned with what people could observe and comprehend. Now, imagine explaining concepts like Wi-Fi, fibre-optic cables, or the intricacies of nuclear fission to someone from that age. They wouldn't merely reject such notions; they might lack the cognitive framework to grasp them entirely.

Fast-forward to today, and our collective "common sense" has evolved. With advances in technology, our understanding of the universe has deepened. As I explored the world and its patterns, both in nature and human behaviour, I began to notice something remarkable. In an age defined by connectivity, platforms like

TikTok and Instagram reveal startling similarities in how we act, think, and feel. Behaviours we might consider unique are echoed by countless others. Why? Are we truly beings of free will, or are we more akin to robots, following pre-programmed patterns?

Today, we pride ourselves on our scientific and technological advancements, yet even now, many of our foundational theories raise more questions than answers. Take the theory of evolution, for instance. While we agree with biologists that genetic evolution has shaped life on Earth, there are gaps in the narrative. Yes, modern organisms differ from their ancient counterparts. But the leap from primitive hominids, such as Australopithecus afarensis .[1]or Sahelanthropus tchadensis.[2], to cognitively advanced humans feels like a story with missing chapters.

The scientific narrative tells us that, over millions of years, these hominids gradually gained cognitive abilities, eventually evolving into us. Yet, I find this explanation hard to accept without question. First, our ancestors, whether we call them hominids or by their more controversial likeness to monkeys, had been living on Earth for millions of years. If evolution was purely a matter of time and opportunity, why didn't other species develop similar cognitive powers? Nature rarely produces one-of-a-kind phenomena, so why are humans the sole species to "win the cognitive lottery"?

[1]An early hominin species that lived about 3.9–2.9 million years ago in East Africa. It is best known for its bipedal locomotion and a mix of human-like and ape-like features, such as a small brain and long arms.

[2] A much older hominin species, dated to about 7 million years ago, discovered in Chad. It may represent one of the earliest ancestors of humans, with a mix of ape-like and hominin traits, including a small braincase and a flat face.

Then there's the issue of interpretation. Scientists have reconstructed entire evolutionary theories from fragments of ancient bones, such as the famous ***Lucy***[3]. But these fragments, millions of years old, leave much room for speculation. While I admire the biologists' effort to piece together a coherent story, I can't help but question how much certainty can truly be drawn from such incomplete evidence. We may look like apes in some ways, and indeed share similarities, but the notion that this alone accounts for the profound mental complexity of humanity feels insufficient.

Moreover, our anatomy raises an uncomfortable question: Were we even designed for this planet? Compared to other species, humans seem oddly fragile. Animals are perfectly suited to their environments, they hunt, adapt, and thrive in the wild. In contrast, we develop cancer under too much sun, suffer plantar fasciitis from walking barefoot, and require filtered water to survive. Could it be that humans are not truly native to Earth, but rather visitors, or constructs, placed here under specific conditions? Much like Sandy Cheeks in SpongeBob Square Pants, we thrive in an artificial bubble within a world that doesn't quite feel like ours.

This thought crystallized for me one day while programming. Staring at the code on my screen, I saw patterns, logic gates and binary decisions that mirrored larger patterns in nature. The realization struck me: could our reality itself be a simulation, a construct governed by an underlying code? The similarities were too profound to dismiss: fractal designs in nature, mathematical constants that define physical laws, and the strange behaviour of particles when observed. Was this evidence of a system designed with precision?

Many would argue that computers are a reflection of the real world. But perhaps the truth is deeper than that. Computers are not built to mimic life; they are built on mathematics. And mathematics, it seems, is the very language of the universe. Perhaps biology can only take us so far in understanding our origins. To truly comprehend where we came from, we might need to turn to mathematics itself, a universal code that binds us, our behaviours, and the cosmos into one interconnected system.

In the end, the recurring patterns, the predictability, and the mathematical elegance of existence may point to something far greater: not merely that we live in a simulation, but that the simulation itself is the essence of what we call reality.

The Da Vinci Code: The Thinker

It was around 400 BCE in Ancient Greece, a time when the air buzzed with philosophical debates and the pursuit of knowledge. Democritus, a philosopher with a curious mind and a penchant for bold ideas, wandered through the bustling marketplace of Abdera. The sun beat down, and as he walked, he observed the dust swirling in the air, illuminated by the light streaming through gaps in the rooftops. He wondered: "What lies between those particles? Could it be... nothing?"

At that time, the prevailing thought, championed by philosophers like Parmenides and Aristotle, was that nature abhorred a vacuum. The very idea of "nothingness" was unthinkable. How could something exist in "nothing"? But Democritus, with his sharp reasoning, dared to question this.

---

[3] The nickname for a famous *Australopithecus afarensis* fossil, discovered in 1974 in Ethiopia. Lucy is approximately **3.2 million years old** and is significant for being one of the most complete early hominin skeletons, providing key insights into bipedalism and human evolution.

Sitting by the shade of an olive tree, he sketched his thoughts in the sand. If matter was made up of tiny, indivisible particles (what he called "atoms"), they must move and interact. For movement to occur, there had to be empty space between the atoms, a void. Without it, everything would be solid, unchanging, and immobile. Thus, the concept of the void was born, though it was purely speculative and challenged the intellectual establishment.

For centuries, the idea of the void languished as a philosophical curiosity. Aristotle's influential rejection of the vacuum dominated Western thought for over a millennium. He argued that nature always filled empty spaces and that the existence of a true vacuum was impossible. The void became a forbidden notion.

Fast forward to the 17th century in Renaissance Europe, where scientific inquiry was starting to challenge ancient dogmas. In Italy, a curious scientist named Evangelista Torricelli, a student of Galileo, set out to explore the properties of air and pressure.

It was 1643, and Torricelli had a revolutionary idea. He filled a long glass tube with mercury, sealed one end, and inverted it into a dish also filled with mercury. To his astonishment, the mercury in the tube didn't completely drain into the dish. Instead, it stopped at a certain height, leaving an empty space at the top of the tube. Torricelli realized this space was devoid of air or any matter, it was a vacuum!

Torricelli's discovery shocked the scientific world. It not only provided the first experimental proof of the vacuum but also showed that air had weight and exerted pressure, a ground-breaking insight.

Torricelli's experiment sparked further investigations. In 1654, Otto von Guericke, a German scientist, constructed the first vacuum pump. He famously demonstrated the power of a vacuum with his Magdeburg hemispheres experiment, where two metal hemispheres, sealed together and evacuated of air, could not be pulled apart even by teams of horses.

From Democritus's philosophical musings under an olive tree to the dramatic experiments of the Renaissance, the void transformed from a radical idea to an undeniable reality. Today, the concept of the vacuum underpins modern physics and technology, from particle accelerators to the vast emptiness of space. What began as an ancient Greek thought experiment has become a cornerstone of our understanding of the universe.

Imagine a child observing the world for the first time, no expensive microscopes, no multimillion-dollar laboratories, no decades of academic study. And yet, with just their innate curiosity, they begin to ask profound questions about how things work. This is the natural gift we all possess: the power of the mind. With logic and imagination, we can observe, theorize, and construct intricate ideas about the world around us. Perhaps even religion, at its core, is a form of philosophy a testament to humanity's capacity to seek meaning.

But we live in an era where this gift is often shackled. Our minds are overwhelmed, chained by distractions that pull us away from deeper thought and focused inquiry. The noise of modern life drowns out the quiet space where ideas once thrived. And even if someone manages to break free, daring to propose a new theory or challenge the status quo, they are swiftly silenced, labelled as a conspiracy theorist, a spreader of disinformation, or a propagandist.

This story reminds us of a crucial truth: the ability to think deeply, to question, and to imagine is both our heritage and our responsibility. It doesn't require vast resources or formal institutions, only the courage to break free from the chains of distraction and the openness to explore without fear of judgment.

As I delved deeper into both religious texts and scientific theories, I found myself caught in a paradox. Religion offered narratives rich with meaning but lacking empirical evidence. Science provided a methodology for understanding the universe but often fell short of answering the *why* behind the *how*. The Big Bang theory, for instance, describes how the universe expanded from an infinitesimal point, but it doesn't satisfactorily explain what set the stage for this monumental event. What force ignited the spark that led to everything we know? With all our technological advancements and accumulated knowledge, why does the fundamental question of our origin remain unanswered?

Reflecting on history, we know that philosophy preceded religion and science. Ancient philosophers grappled with existential questions long before organized religions provided more accessible narratives. They pondered the essence of being, the fabric of reality, and the mechanics of the cosmos. However, their ideas were often esoteric, accessible only to those with the leisure and education to understand them.

Religion, in many ways, can be seen as a bridge translating complex metaphysical concepts into stories and doctrines that the masses could grasp. It's akin to explaining the intricate operations of a computer by simply saying, "When you click this button, this happens," without delving into the underlying code and electrical impulses.

Science, however, speaks in the language of the universe, numbers rather than words, a language that can be elusive for most to grasp. While we attempt to interpret it with words, the translation is never quite the same; only those fluent in numbers truly understand the universe's essence. This disconnect is part of why the scientific narrative of existence often lacks the widespread resonance of religious or philosophical tales.

As previously discussed, we inhabit a universe defined by recurring patterns. What has happened once can, and often does, happen again. This principle is not just a quirk of nature but a key to unravelling the unknown. When confronted with a question that seems beyond your grasp, the answer might already be embedded in the world around you. Observe how systems function, delve into the thoughts of those who sought similar answers in the past, and explore the wealth of knowledge that science has to offer. In doing so, your mind will naturally begin to generate possibilities. Among these, there is a high likelihood that at least one will align closely, if not perfectly, with the underlying truth.

Open Your Mind: Imposs/able

Venturing into subjects that challenge conventional wisdom requires an open mind and a willingness to embrace uncertainty and entertain possibilities that lie beyond accepted norms. It's easy to dismiss unproven theories, but doing so closes the door to potential breakthroughs.

History is replete with examples of ridiculed or suppressed ideas, only to be validated later. Galileo Galilei faced persecution for supporting heliocentrism, yet his ideas eventually revolutionized our understanding of the solar system.

Today, we live under vast, super-controlling institutions where ideas that challenge their authority or systems are quickly labelled as outlandish, subversive and conspiracy theories.

Consider Operation Mockingbird[4]: once dismissed as conspiracy, now acknowledged as fact. Imagine if someone introduced the concept of a deity in the sky today; it would be branded as conspiracy, misinformation, or outright banned. This resistance to new perspectives isn't new; what was once the domain of the Church now manifests as government, legislation, and legal prosecution.

Looking back, we see figures like Leucippus and Democritus, who dared to imagine atoms in a time devoid of evidence or tools to prove their existence. It was through imagination that they solved the mysteries of matter. Imagination remains one of our most powerful tools, and many scientific truths began as philosophical musings before they were ever proven. Yet, we are moving towards a future where certain entities, whoever they are, seem intent on restricting the space for free thought, demarcating what's "acceptable", and silencing any deviation from it.
Who would have thought education could be weaponized? Where lack of education once implied ignorance, the modern education system increasingly declares rigid absolutes. Anything beyond those boundaries is branded wrong and opposed. This might seem minor, but it's a chain on the mind, stifling the growth of free thinkers and critical analysis. It is essential to reclaim the power to assess, decide, and define our perspectives, not merely accept the boundaries laid out for us.

As I reflect on the world today, it's tempting to believe that the eras of slavery, colonialism, and imperialism are relics of the past, long buried in history. But that's not true. None of these constructs have disappeared, they've simply evolved into more subtle, modern forms. They've shifted into frameworks that go unrecognized as neo-colonialism by those who control the narrative. Consider this: there are governments today that travel across the globe, not with chains or guns, but with ideas, working to reshape how others think, to normalize certain lifestyles, and to convince entire populations that their beliefs, values, and ways of life are wrong. And what, then, is the "correct" way? The one dictated by the neo-colonialism themselves.
Think about why education is mandatory in these so-called modern nations. Schools don't just teach skills; they craft narratives, moulding the perspectives of their citizens from an early age. One could argue that this isn't inherently bad, it fosters societal cohesion and perhaps even global harmony if everyone agrees on certain universal principles. But herein lies the problem. This ideological imperialism, this subtle enforcement of a singular worldview, suffocates the very essence of free thought. It cripples the ability to question, to think independently, to dream beyond the bounds of what is deemed acceptable.

What happens to those who dare to dissent? They find themselves surrounded by individuals who behave like programmed machines, robots of the mind, unwilling or unable to entertain alternative perspectives. Would you dare to voice a belief that challenges their entrenched narratives?

To me, it feels as though we're living in a world populated by followers, by spectators, by what some might call NPCs, non-player characters in a grand, orchestrated simulation. These individuals adopt beliefs as they're told, live according to whatever trends dominate, and allow their opinions to align seamlessly with the majority. Nothing stands out. They are

---

[4] Operation Mockingbird was a covert program initiated by the CIA in the late 1940s to influence media and control public opinion during the Cold War. Through this operation, the CIA recruited journalists and planted stories in major news outlets to promote narratives favourable to U.S. interests and counter Soviet propaganda. The operation involved prominent media figures and organizations, providing the CIA with a powerful tool to shape public perception domestically and internationally. The program was exposed in the 1970s, leading to scrutiny of government influence in the media.

the perfect embodiment of "2 + 2 = 4," a predictable equation devoid of depth, mystery, or creativity.

But we weren't made to be this way. Humanity is meant to be unpredictable, a tapestry woven with free will, diverse opinions, and countless perspectives. We're supposed to challenge, to question, to explore the uncomfortable edges of our understanding. Yet, here we are, in a world that celebrates conformity and stifles individuality. And that, more than anything else, feels like the greatest loss of all.

Here is an interesting question: even if you're open to the idea that our reality may not be the primary one, would your mind allow you to entertain it? Scientifically and psychologically, this process is far more complex than a simple yes or no. Our brains are wired to seek patterns, establish consistency, and lean on "common sense".[5] to simplify the world. This reliance on mental shortcuts is an evolutionary mechanism, conserving cognitive energy by favouring familiar perspectives and often dismissing novel or challenging ideas.

When confronted with disruptive concepts, like the possibility of a simulated reality, the brain instinctively resists. Embracing such ideas would mean unsettling these shortcuts, leading to cognitive dissonance, a mental discomfort arising from conflicting beliefs. Accepting a perspective that challenges "common sense" demands a taxing re-evaluation of foundational assumptions, which the brain resists to conserve energy. Beyond cognitive effort, a shift in worldview can threaten social bonds and identity, as our sense of belonging often hinges on shared beliefs. Thus, by resisting unfamiliar ideas, the brain preserves not only cognitive efficiency but also social harmony.

Our minds have evolved mechanisms to manage the anxiety of uncertainty. When unsure, we often reduce dissonance by convincing ourselves that our chosen answer is correct. Cognitive biases like commitment and consistency bias drive us to stick with decisions to avoid the discomfort of re-evaluating them. Heuristics, anchoring, and overconfidence bias create a sense of certainty, even when we lack complete information, helping us function effectively amid ambiguity. These mechanisms work in tandem to maintain stability, enabling us to navigate life without being overwhelmed by doubt.

This interplay of cognitive efficiency, social bonds, and emotional stability acts as a mental safeguard, often compelling us to adhere to familiar narratives. When faced with new perspectives challenging established beliefs, like the notion that reality might be simulated, the brain employs defence mechanisms to resist unsettling ideas. A common first line of defence is discrediting the idea entirely. The brain may dismiss such concepts as impractical or meaningless, steering us away from deeper consideration. It may also cast doubt on the credibility of information that contradicts our current beliefs, trivializing or questioning the source to avoid engaging with the idea fully.

Confronted with conflicting information, the mind quickly seeks to reduce cognitive dissonance. Often, it achieves this by rationalizing the existing belief, subtly persuading itself that the disruptive thought is incorrect or irrelevant. Heuristic processing enables swift dismissal, "it doesn't fit with what I know, so it must be wrong", while anchoring biases us toward established beliefs, making it challenging to adopt new perspectives.

---

[5] Common sense refers to the intuitive knowledge and beliefs that people develop through experience, social learning, and cultural influence. It's the set of assumptions we hold about how the world works, guiding our expectations, judgments, and actions. Common sense helps us navigate everyday life without needing to analyse every detail deeply; it offers shortcuts for interpreting situations, allowing us to make quick, reasonably accurate decisions. However, common sense can be limiting, as it's based on familiar ideas rather than objective or critical analysis.

Overconfidence can further reinforce resistance, giving us a false sense of certainty in our current understanding. The mind set of "I already understand reality; there's no need to question it" shuts out alternatives. Social validation amplifies this, as we feel secure holding views that align with those around us. When a new idea threatens this harmony, the brain resists to maintain the comfort of shared perspectives and avoid potential isolation.

Emotions play a role, too, as challenging ideas often provoke discomfort, fear, or confusion. The brain instinctively avoids unsettling thoughts, using emotions as deterrents. Once we've committed to a belief, consistency bias discourages change, reinforcing our comfort and stability.

Finally, practicality often prevails; disruptive ideas are easily dismissed as irrelevant to daily life. By prioritizing familiar, "practical" concerns, the brain preserves cognitive stability, social cohesion, and emotional equilibrium, maintaining order at the cost of limiting openness to profound insights. Through mechanisms like discrediting, anchoring, rationalizing, and appealing to practicality, the mind conserves energy, reinforces social harmony, and sustains cognitive stability, often at the expense of exploring alternative realities.

The human mind, in many ways, resembles a robot's programming, bound by constraints and conditioned responses that seem to dictate how we process reality. Just as robots are programmed with algorithms that determine their responses and limit their capacity to act outside those parameters, our minds operate with mental shortcuts and biases that streamline thinking but restrict our ability to explore certain perspectives fully. This cognitive programming makes us lean toward established beliefs, often filtering out ideas that could disrupt our perception of reality, like questioning if this reality itself is primary. Much like a robot whose code prevents it from thinking beyond its set tasks, our mental "programming" keeps us within a comfortable, predictable loop. It's as if our brains have been conditioned to avoid thoughts that challenge the status quo, creating a built-in resistance to ideas that might destabilize our understanding of the world and, in doing so, safeguarding mental stability at the expense of profound curiosity.

Those who think differently from the majority, who question reality or resist common sense, are often labelled as "mentally ill," "crazy," or "delusional." This reaction isn't just a social reflex but a deeply ingrained psychological response rooted in the brain's desire for stability and consistency. When someone deviates from widely accepted beliefs, they disrupt the shared "programming" that gives society its sense of order. Our brains are wired to view unusual perspectives as a potential threat, as they challenge the security of a predictable world. As a result, society often marginalizes individuals with unconventional ideas, framing them as mentally unstable to preserve the collective comfort of shared beliefs. This labelling acts as a defence mechanism, reassuring the majority that their worldview is valid and shielding them from the discomfort of having to question their own assumptions. Those who see beyond the ordinary are not necessarily "crazy" but are simply unbound by the mental shortcuts that keep most people anchored in a shared version of reality, and this divergence can be both unsettling and misunderstood.

How did our brains develop such a strong defence mechanism against challenging beliefs? If we lean on the theory of evolution, humans, as an ancient species, millions of years old, have undergone countless adaptations. Over time, our brains appear to have honed this defensive mechanism to such an extent that, today, many people reject factual reality outright. Rather than confronting inconvenient truths, they often choose the comfort of self-deception. If we accept the premise that humans were not originally equipped with these mental defence systems and developed them gradually, we must ask: does this evolution signify a loss of cognitive power or a gain in adaptability?

So, how do we challenge this "robotic" mode of thinking? A full exploration of this question could fill another book, but in simple terms, the key lies in embracing uncertainty. Introduce a "maybe" to everything you believe, hear, or even know. Did the USA land on the moon in 1969? Maybe. Is coffee good for you? Maybe. Does 2+2 always equal 4? Maybe. Is your biological father really your father? Maybe, Is *Skibidi* Sigma? Maybe. Why this approach? Because we exist in a world of superposition[6], where truths are often layered, conditional, and not as absolute as they seem. Accepting anything as definitive truth limits the brain's ability to explore the spectrum of possibilities.

By questioning even the most basic assumptions, you begin to loosen the mental constraints that bind your thinking. Truth becomes a spectrum, and it's up to you to discern what is logical and resonates with deeper inquiry. This simple act, adding a "maybe", empowers the mind to push beyond its safety mechanisms, allowing it to think freely and explore ideas with greater depth and clarity.

There is a Persian proverb that says, "They asked the fox, 'Who is your witness?' and he said, 'My tail.'" This reminds us that, like a fox trusting its own tail, we often rely on our own biased perceptions as proof of truth. In the same way that machine learning models are programmed to think and decide based on input data and training, our minds have been programmed, not only since birth but also across generations. According to epigenetics[7], experiences, thoughts, and even emotional states can be passed from one generation to the next.

This means that your mind is not a blank slate at birth; it carries the imprints of the past, including biases encoded in your DNA. Fundamentally, this suggests that even your logic, the tool you use to discern truth, is inherently biased. When you question the foundation of your existence or consider the nature of reality itself, you are not just questioning ideas but battling millions of years of mental programming designed to protect and preserve a specific understanding of the world. Perhaps it is this deeply ingrained bias that makes the concept of questioning reality seem absurd or even laughable to the programmed mind. Yet, this is precisely why embracing uncertainty is so crucial, it allows us to push against the weight of inherited assumptions and explore the unknown.

Evidence: No Body, No Case.

Often, we speak of the nature of reality, whether our lives unfold in some grand cosmic theatre or an elaborate simulation meticulously hiding its own seams. People search relentlessly for "glitches," anomalies that betray the artificial scaffolding of our existence. Yet, ironically, the human biological framework is wired to fear precisely the sort of disruption that a genuine glitch would cause. Our brains are masters of pattern recognition; it is how we have survived predators and natural catastrophes for millennia. Should something appear that blatantly violates our understood patterns, like a portal ripping open in your living room, it would not merely shock your senses but potentially overload your cardiovascular system to a fatal degree. The primal fear centres in your brain would fire incessantly, spiking adrenaline and cortisol, and quite possibly stopping your heart before you could even decide whether to run.

---

[6] **Superposition** is a fundamental principle of quantum mechanics. It states that a quantum system, like an electron or a photon, can exist in multiple states at once until it is observed or measured.

[7] **Epigenetics** is the study of changes in gene expression or activity that do not involve alterations to the underlying DNA sequence. These changes are influenced by environmental factors, lifestyle, and experiences and can affect how genes are turned on or off.

But what if you miraculously survived witnessing such an event? The tragedy is that your account would likely be dismissed as a delusion, relegating you to the realm of those deemed "mentally ill." After all, modern society leans heavily upon empiricism, no proof, no truth. We label as conspiracy theorists or psychotic anyone who claims to have witnessed the impossible. This framework of scepticism may serve to safeguard us from charlatans, but it may also blind us to truths that cannot be easily tested with the scientific apparatus at hand. Consider the skewed logic: if your heart raced uncontrollably and you fainted during your 'vision,' doctors might interpret it as a medical anomaly or mental break, never once considering your story might have credence. It is a cruel paradox: the moment you see something that could unshackle us from our comfortable illusions, you become the very outcast whose words are discredited and dismissed.

Interestingly, this double standard extends beyond paranormal phenomena. We live in a world where many accept extraordinary claims with minimal or no evidence, be it a divine overseer meticulously documenting every human action or the belief in intangible energies and 'miracle' remedies, yet balk in fierce denial at other claims that are arguably no more fantastical. History offers many cautionary tales: think of how opium was pushed onto societies during the Opium Wars.[8], systematically weakening a population for political advantage. Or consider how rum and other intoxicants were historically used to manipulate indigenous cultures, subduing them for imperialistic gain. Despite these well-documented manipulations, people still roll their eyes when told that a commonly used, seemingly benign substance, alcohol, nicotine, even marijuana, could be part of a larger strategy to keep us sedate and malleable. It is deemed "ridiculous" or "paranoid," even though it aligns uncomfortably well with documented precedents.

This calls into question our entire metric for truth. Some dismiss anecdotal observations outright, yet place unwavering faith in intangible entities or theories for which no direct proof has ever been provided. Does the absence of proof indeed prove non-existence? Consider the thought experiment of the falling tree in a silent forest. Science usually insists, "Yes, it still makes a sound," thanks to the laws of physics that would apply whether we're there or not. But from the viewpoint of pure phenomenology, "sound" only exists within the ear of a listener. Both perspectives hold elements of truth, highlighting that reality can hinge on context and interpretation.

This same interpretative dilemma arises in socio-political events. Take, for example, the claims of election fraud in the 2016 U.S. presidential race. Despite repeated investigations, no definitive evidence emerged to universally confirm or debunk the accusations. To some, the lack of proof is evidence of innocence; to others, it is proof of a cunning cover-up. The truth remains in a liminal space where partisans latch onto whichever narrative they prefer. The principle is the same as discrediting a friend through rumours and lies. Once the friend's reputation is ruined, any protestations they make are written off as the ramblings of a disgraced individual. Thus, when a real injustice finally does befall them, no one believes a word they say.

---

[8] The Opium Wars were two conflicts in the mid-19th century between China and Western powers, primarily Britain. The First Opium War (1839–1842) erupted when China tried to stop British traders from smuggling opium into the country, which had led to widespread addiction and social issues. Britain, seeking to protect its lucrative opium trade, defeated China, leading to the Treaty of Nanking. This treaty forced China to cede Hong Kong to Britain and open several ports to foreign trade. The Second Opium War (1856–1860), involving Britain and France against China, further weakened Chinese sovereignty, resulting in more open ports and foreign control over trade. The wars marked the beginning of the "Century of Humiliation" for China, with Western powers exerting control over Chinese territory and economy.

Likewise, the simulation hypothesis teeters on this same precipice. On one side, we lack definitive evidence of a cosmic supercomputer generating every quantum fluctuation. On the other, we have nothing definitive to disprove it either. Sceptics scoff that a simulated reality is a bridge too far, yet ironically accept premises like an omnipotent being orchestrating every facet of existence. If our reality were a simulation, designed to be self-sealing and resistant to detection, then its very perfection would explain why we lack conclusive "glitch" evidence. Just because our instruments detect no seams doesn't mean those seams aren't deliberately and masterfully hidden.

Our ignorance, in that sense, might even be a critical failsafe preventing existential panic. Indeed, if we knew for a fact that we were fish confined to a high-tech tank, many would lose their will to strive, to innovate, or to act ethically. And here is the ultimate irony: perhaps not knowing is what makes life richer and more bearable. We move through existence enlivened by the possibility of mystery, driven by curiosity to explore nature's intricacies, social dynamics, and the bounds of our own consciousness. Once the bright line of absolute certainty is revealed, once you realize that your entire world is no more than a line of code or that your next-door neighbour truly did see an alien stepping out of a rip in space-time, our entire concept of reality shifts, often dangerously. The fish may panic upon realizing the glass walls around them are not infinite ocean but a restricted domain crafted by a distant observer. That delirium could unravel individuals, and, by extension, entire societies.

Hence, searching for undeniable glitches might be the least productive quest of all: if you found one, you may not survive its psychological or physiological fallout; if you survived, no one would believe you; and if your community finally did believe you, it could catapult civilization into nihilistic chaos. Yet, refusing to contemplate these possibilities is equally limiting. Our challenge, then, is to remain open to the improbable while resisting blind acceptance of extraordinary claims. It is a delicate balance, but maybe that's the point of our predicament. We are forced to navigate this ephemeral plane with a curious mind and a guarded heart, ever aware that the line between truth and illusion is precariously thin, ready at any moment to rewrite our sense of the real.

The Language of the Universe: Numbers

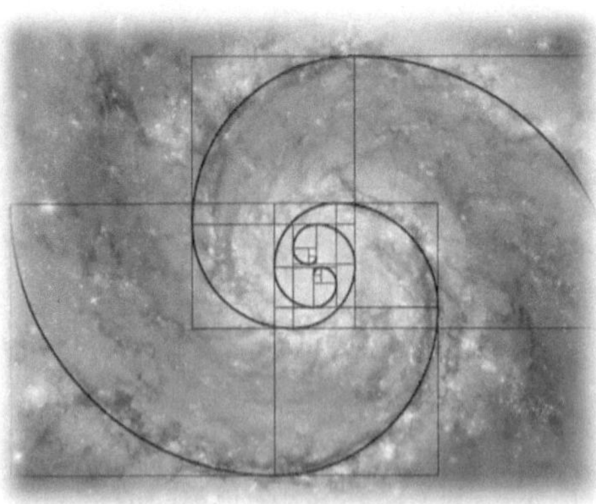

Numbers are not mere symbols on a page; they are the cosmic choreography that underpins our very existence. As far back as ancient civilizations, priests and philosophers alike revered numbers as divine keys that could unlock the secrets of being. Today, some of us still cling to the illusion that our eyes and instruments are the best windows into reality. But if we dare to look deeper, to the raw and unyielding certainty of mathematics, we find a universe thrumming with numerical patterns that shape and propel every atom, every galaxy, and every flicker of consciousness.

Consider, for instance, the Fibonacci sequence, that seemingly simple progression of 0, 1, 1, 2, 3, 5, 8, 13, and so on. Unremarkable at first glance, it swiftly reveals its fingerprints throughout nature, from the spiralling seeds of a sunflower to the branching patterns of trees. These spirals and ratios aren't frivolous quirks; they are elegant solutions, arrived at through cosmic trial and error, optimizing everything from the packing of seeds to the arrangement of leaves around a stem (phyllotaxis). And it doesn't end there. Delve deeper into the whispering geometry of nautilus shells or the curved horns of rams, and again, you'll see the same numerical tapestry re-emerging, as though reality is screaming, "Here is the code."

This is where the golden ratio, commonly denoted by the Greek letter phi (φ), arises like a sacred touchstone of nature. Closely tied to Fibonacci's progression, the golden ratio emerges in art, architecture, and the swirling structure of entire galaxies. In spiral shells, flower petals, and even our own anatomy, φ surfaces again and again, reinforcing the eerie sense that our world is less physical than it is mathematical. And while phi captures the poetry of spirals, pi (π) reigns as the emblem of perfect cycles, circles that define celestial orbits, wave patterns, and the subtle arcs of quantum fields. From the swirl of water down a drain to the path planets trace around stars, π is woven into the dance of cosmic motion.

If Fibonacci, phi, and π prove anything, it's that mathematics is not just an abstract pastime but the blueprint of existence. Galaxies, plants, animals, even subatomic particles, all follow numerical laws that transcend our ordinary senses. We often limit ourselves by relying solely on observation, thinking we must "see to believe." Yet the most vital truths of our universe, the quantum leaps in electron orbitals or the fusion hearts of distant stars, are visible only through equations and constants. Numbers let us peer into the past, project ourselves into the future, and bridge unimaginable scales, from the microscopically small to the cosmically vast.

Look to the heavens if you doubt the reality of these patterns. The spiral arms of the Milky Way bear a likeness to Fibonacci's sequence, shaped by the gravitational ballet of countless stars. On Earth, entire ecosystems revolve around recurring numerical rhythms: the cycles of tides, the migrations of animals, and the beating of our own hearts, each meticulously in step with the math that undergirds them. Some may shrug these off as cosmic coincidences, but what if they are not coincidences at all? What if the nature of reality is nothing but an endless repetition of numeric truths, coded into the very fabric of space-time?

To grasp this is to realize that mathematics is not just the language of the universe but its metaphysical bedrock, the purest expression of truth upon which all else is built. In a radical sense, numbers might be more "real" than we are. We come and go, but φ, π, and the Fibonacci sequence persist, guiding the formation of stars, the geometry of galaxies, and the swirling complexities of life. The audacious claim is that to truly know the essence of reality, we must turn from the flickering shadows of our senses and devote ourselves to the luminous certainty of numbers. Where eyes fail, equations succeed. Where instruments fall silent, constants sing.

Embracing this perspective means abandoning the comfort of purely visual understanding and acknowledging a far more controversial possibility: that our physical reality is but an emergent property of a deeper, numerical domain. We can either scoff or stand in awe at this suggestion. But as history shows, wherever mathematics has probed, the universe eventually yields its secrets, sometimes with shocking familiarity, sometimes with radical surprises. It is the mathematician's pursuit that has cracked the cosmic safe time and again, not the fleeting illusions of human sight.

So what is reality? A dream in the mind of a cosmic mathematician? A grand tapestry of equations existing beyond time and space? Perhaps both, perhaps neither. One thing is certain: if we wish to unlock the deepest truths of existence, we must learn to read this ancient code, scrawled in Fibonacci numbers, guided by phi, haunted by pi, and orchestrated by an unyielding logic that hums through every corner of creation. Only then can we begin to glimpse the architecture of reality, bold, endless, and unapologetically numerical.

The Quantum Illusion of reality: Why?

In writing this book, my aim is not to convince you of any particular viewpoint but to invite you on a journey of exploration. Together, we'll traverse the landscapes of science, philosophy, religion, and technology. We'll question the nature of reality, consider the limitations of human perception, and ponder the possibilities that emerge when we dare to think beyond the conventional. This book is an invitation to open your mind, challenge your beliefs, and engage with ideas that might initially seem improbable but could hold the key to unravelling the mysteries of our existence.

You might ask why. Why is there so much information, scientific data, and philosophical discussion in this book? Why do we delve into topics like gravity when we're contemplating whether our reality is the primary reality, or perhaps just a copy of a primary one? The answer lies in the nature of the question itself. We are probably never going to factually prove that our reality is not the primary one. That kind of certainty lies beyond the capabilities of the human mind; the concept is simply outside the bounds of our perception and understanding.

However, what we can do is examine the evidence and ideas that surround this subject. Since this question touches on the existence of the entire universe, it requires us to consider a vast array of information. To make an informed judgment, you need a deep understanding of how things work within this universe and within this reality. It's not just about entertaining abstract theories; it's about grounding those theories in the principles and mechanics of the universe we inhabit.

Every significant advancement in human thought began with a question that challenged the status quo. The subject of this book is no different. It demands that we step outside our comfort zones, that we engage with science, philosophy, and the unknown with equal intensity. This knowledge equips you to be more than a passive observer. It empowers you to be the judge.

Ultimately, it is you who will decide. Is this reality the primary one, or merely a shadow of something greater? The answers might not come easily, and they may never arrive with absolute certainty. But the pursuit of understanding itself is profoundly worthwhile. After all, in this relentless search for knowledge, we move closer to the truth, whatever that truth may ultimately be.

So, I encourage you to read on with an open heart and an inquisitive mind. Together, let us embrace the unknown, question the foundations of reality, and acknowledge that sometimes the most profound answers are not found in certainty but in the journey toward understanding. Let's embark on this intellectual adventure, fuelled by imagination, scepticism, and the boundless curiosity that defines us as human beings.

Foundation

Reality and non-reality are threads of the same fabric, entangled in a dance where each defines the other.

What is reality?

Reality is often assumed to be the bedrock of our existence, the one immutable constant in an otherwise chaotic universe. We go about our lives trusting what our senses tell us: the solidity of a table, the warmth of the sun, the sound of our own breath. These experiences collectively reassure us that there is, indeed, "something" out there. Yet, when we venture deeper, we find that the concept we hold so dear, this matrix of tangible objects and familiar sensations, may itself be a shadow projection of a far more complex truth. What if our most fundamental assumptions of what's "real" are merely the floor of a stage set, and the actual stage extends infinitely in all directions?

When we talk about a possible "simulation," it's not an attempt to dismiss the realness of everyday life. A simulated reality can still be entirely convincing to those who inhabit it; pain still hurts, food is still nourishing, heartbreak still scars. Yet one crucial difference persists: there can be only one ultimate reality, and by definition, if we find ourselves in a created or layered environment, we do not dwell in that ultimate realm. Like characters in a meticulously designed video game, our experiences remain meaningful and profound to us, even if the "console" generating them exists on a higher level we cannot directly perceive.

Think of the game characters in a digital landscape, bound to the contours of their grid and the rules of their coded physics. They have no way of stepping outside that code to see the programmers behind the curtain. Should they discover their simulation, that revelation wouldn't negate the authenticity of their emotional or physical experiences. They would still run from danger, feel a rush of excitement, and fear the unknown. Everything remains real to them, but not real in the sense of being the primary or cosmic baseline.

So, too, might we be bound to a three-dimensional environment, limited by senses that have evolved only to keep us alive in a world that might be an infinitesimal fraction of a grander multidimensional cosmos. We break down distance into miles, measure time in seconds, and rely on eyes that can only perceive a sliver of the electromagnetic spectrum. If our existence is part of a vastly more intricate framework, how could we, as three-dimensional beings, possibly step into the code itself and see the full architecture?

Controversially, some theories suggest we are indeed far from the bedrock of reality. Instead, we might occupy a carefully orchestrated environment, an "island universe" within something infinitely larger. To us, this environment is not an illusion. It's the only environment we know; it shapes every reference point we have. But it may still be akin to training wheels, a constructed scaffolding meant to contain our existence and our perception, while the grand "source code," or actual reality, lies beyond our comprehension.

Modern science is rife with clues that we are missing some fundamental layer of truth: quantum phenomena defy classical logic, consciousness remains a mystery, and the computational leaps we're achieving suggest that reality could be processed information. If so, it might not be long before we ourselves become capable of designing such immersive simulations. It may turn out that creating entire realities is just another step on the cosmic ladder. And if we can do it, who is to say it hasn't been done before?

Whether you believe we are part of a cosmic sandbox, a divine creation, or the original reality itself, one point remains central: there cannot be two equally fundamental "primary" realities. The claims of simulation theorists do not trivialize our experiences; our joys and sorrows still leave tangible echoes in our minds. Rather, they challenge us to redefine what "primary" means. If that bigger truth exists, then this life, while utterly real to us, may be a nested simulation, beautiful, vivid, and profound, yet not the final rung on reality's ladder.

In one sense, none of this diminishes the value or authenticity of our day-to-day existence; in another sense, it obliterates the concept of permanence we cling to. We might be akin to characters in a dream, enthralled by self-contained stories that, in their own context, carry immense weight. But if the dreamer awakens, the reality shifts, and the dream's significance mutates. So too might our own reality be recalibrated if we peel back the layers, perhaps to a dimension with more than three axes of space and a fundamentally different logic of time. In short, we might have to face the disturbing possibility that our notion of reality is just a fraction of the puzzle, a shadow cast on the wall of an even vaster cosmos.

Will we ever be able to transcend our limits and see the scaffolding for what it is? Or are we like goldfish in a bowl, gazing out through curved glass, never glimpsing the full room beyond the water? One thing is certain: for every answer we stumble upon, new questions arise, each one tugging us deeper into the grand enigma. If there truly is a primary reality existing beyond our own, then our struggles and triumphs are stages in a cosmic drama that has only just begun to unfold. And perhaps that knowledge can embolden us, knowing that within the very constraints of our "simulation," we have found curiosity and wonder, forging a path that may one day guide us home to whatever "home" truly is.

As I dived deeper into philosophy and science, I was drawn to two competing worldviews about the nature of reality: Materialism and Idealism. They were not just academic theories but reflections of humanity's deepest attempts to understand existence.

Materialism

Materialism, a philosophical doctrine asserting that physical matter is the fundamental substance in nature, has its roots deeply embedded in ancient Greek philosophy. Traced back to the ideas of Democritus, materialism emerges from his vision of the universe as a vast, intricate dance of indivisible atoms moving ceaselessly through the void. These atoms, varying in shape, size, and motion, combined to form all the phenomena observed in the world. Democritus, along with his mentor Leucippus, introduced this atomistic view as a means of explaining natural processes without recourse to supernatural or divine intervention, a radical departure from the mythological explanations of his time. This early materialism provided the groundwork for later debates on the nature of reality, emphasizing that everything, including the mind and soul, could be understood in terms of physical processes.

The legacy of materialism extended into the Roman era through the works of Epicurus and his follower, Lucretius. Epicurus adopted and modified the atomistic framework, proposing that the random "swerve" of atoms introduced free will into an otherwise deterministic universe. Lucretius' epic poem *De Rerum Natura* (On the Nature of Things) vividly illustrated these ideas, blending poetic eloquence with philosophical rigor to argue that the material world is governed by natural laws, not the whims of capricious gods. This articulation of materialism reinforced the notion that sensory experience and empirical observation were crucial for understanding reality, a perspective that would later resonate in the scientific revolutions of the modern era.

During the Enlightenment[9], materialism underwent significant transformation as it intersected with the burgeoning scientific discoveries of the 17th and 18th centuries. Thinkers like Thomas Hobbes advanced a mechanistic view of the universe, positing that all phenomena, including human thought and emotion, could be explained in terms of matter and motion. This mechanistic materialism found echoes in the works of French philosophers like Julien Offray de La Mettrie, whose *L'Homme Machine* (Man a Machine) argued provocatively that humans were complex biological machines. Such views challenged traditional religious doctrines, provoking sharp debates over the role of materialism in understanding the nature of existence and the human condition.

The 19th century witnessed a deepening of materialist thought, particularly in response to advancements in biology, physics, and chemistry. The rise of dialectical materialism, as articulated by Karl Marx and Friedrich Engels, marked a significant evolution of the doctrine. Influenced by Hegelian dialectics and scientific materialism, Marx and Engels argued that material conditions and economic relations shaped human societies and historical development. This materialist conception of history emphasized the interplay between human labour, class struggle, and the means of production, forming the theoretical backbone of Marxist philosophy and socialist movements.

In my reading, I stumbled upon an intriguing connection: the relentless human pursuit of unifying principles that transcend individual disciplines. This endeavour finds its roots in both mathematics and physics, where thinkers have sought a single, elegant framework to explain the complexities of the universe. Leonhard Euler, one of history's most brilliant mathematicians, made a ground-breaking contribution to this quest in 1734 when he introduced an equation that seamlessly encompassed all the fundamental constants of mathematics. Known today as Euler's identity, this equation, $\boldsymbol{e^{i\pi} + 1 = 0}$, is often referred to as **"the God equation."**

It astonishingly connects five fundamental mathematical constants, **e** (the base of natural logarithms), **i** (the imaginary unit), **π** (the ratio of a circle's circumference to its diameter), **1**, and **0**, in a single, harmonious expression. Mathematicians and physicists alike have marvelled at its beauty, describing it as a glimpse into the deep structure of reality.

Decades later, Albert Einstein, a towering figure in physics, drew inspiration from this same kind of mathematical elegance. His work on the theory of general relativity revolutionized our understanding of space, time, and gravity. Yet, even as he unlocked some of the universe's secrets, Einstein's ultimate goal remained elusive: he sought a single, overarching equation that would unify the forces of nature and provide a comprehensive explanation of reality itself. This quest, often referred to as the search for a **"Theory of Everything,"** captivated Einstein for over 30 years, driving him to explore the unification of gravity, electromagnetism, and the nuclear forces.

Einstein's efforts were part of a broader narrative in science, deeply rooted in materialism. Materialism gained significant traction during the modern scientific era, bolstered by ground-breaking advancements. Isaac Newton's laws of motion depicted the universe as a deterministic, clockwork system governed by precise mathematical laws. Charles Darwin's theory of evolution provided a material explanation for the diversity of life, emphasizing

[9] The Enlightenment, often referred to as the "Age of Reason," was an intellectual and cultural movement in 17th and 18th-century Europe emphasizing reason, science, and individualism over tradition and religious authority. Thinkers of the Enlightenment sought to reform society using rational thought and empirical evidence, influencing fields such as politics, philosophy, and art. The movement's ideals laid the foundation for modern democracy, human rights, and secular governance.

natural selection over divine intervention. Even Einstein's early work reinforced the materialist framework, describing a cosmos governed by predictable, measurable forces.

However, as Einstein's search for a unifying theory faltered, he began to question the very foundations of materialism. His reflections grew increasingly philosophical, grappling with the mysteries of time and the atomistic nature of matter. In 1954, Einstein candidly remarked:

> *"I must confess that I was unable to find an explanation for the atomistic character of nature. One must find a way to avoid the space-time continuum altogether, though I have not the slightest idea what kind of elementary concept could be used in such a theory. The distinction between past, present, and future is only a stubbornly persistent illusion, however tenacious this illusion may be."*

These words reflect a profound shift in Einstein's perspective. Once a champion of materialism, he now doubted its sufficiency to explain the deeper truths of existence. His disillusionment was not an isolated phenomenon but mirrored broader philosophical trends. Materialism, while formidable in its scientific applications, has increasingly faced challenges from idealism, a worldview that posits consciousness, rather than matter, as the fundamental substance of reality.

Quantum mechanics, with its strange, observer-dependent phenomena, has further fuelled this shift. Experiments revealing the role of observation in determining physical outcomes suggest that the fabric of reality might be shaped by non-material forces, or consciousness itself. These insights have led some to propose that the universe is not merely governed by mathematical laws but could be a manifestation of something more profound, perhaps even a product of consciousness or a higher-order principle.

This shift raises profound questions about the nature of reality. If the universe is fundamentally mathematical, as hinted by Euler's identity and Einstein's quest for unification, is mathematics itself the ultimate reality? Or does it point to something even deeper, something that transcends both materialism and mathematics?

Today, the agenda appears to be shifting. Materialism, which once seemed poised to explain the entirety of existence, is now facing a resurgence of idealist thought in both philosophical and scientific circles. As we unravel the mysteries of quantum mechanics and delve into the philosophical implications of our discoveries, we find ourselves standing at the threshold of a new paradigm, one that challenges our understanding of the universe and invites us to explore the profound connection between matter, mathematics, and consciousness.

Idealism

As I contrasted materialism with idealism, I found myself stepping into an entirely different world, a realm where reality was no longer constrained by physical substance. Idealism, championed by Plato, argued that the physical world was merely a shadow of a higher, intangible reality. To Plato, the world I touched, saw, and experienced was not truly real; it was but a pale reflection of a more perfect realm of ideas, accessible only through intellectual contemplation. His allegory of the cave vividly illustrates this perspective, depicting humanity as prisoners mistaking shadows on a wall for the entirety of existence, while the true forms, unchanging, eternal, and perfect, reside in a higher plane.

Centuries later, idealism evolved through the work of philosophers like George Berkeley, who took Plato's concept to its logical extreme. Berkeley argued that matter did not exist independently; instead, objects existed only when perceived by a conscious observer. According to Berkeley, the universe persisted because a divine observer, God, was ever-present, ensuring continuity. This radical claim, that the existence of the physical world depended entirely on perception, challenged the bedrock of materialism and foreshadowed some of the most perplexing findings in modern science. Einstein's later reflections, though not strictly idealist, echoed some of these themes in surprising ways.

His statement that

"The distinction between past, present, and future is only a stubbornly persistent illusion"

Aligns with Plato's notion of a higher truth beyond physical appearances. Einstein's disillusionment with the materialist framework paralleled the philosophical pivot from a purely mechanistic universe to one that hinted at deeper, intangible realities.

Quantum mechanics, emerging in the early 20th century, added new layers of complexity to the debate. I read with fascination about the double-slit experiment, where particles seemed to "choose" their behaviour, acting as waves or particles, based on whether they were observed. This experiment, at its core, suggested an unsettling idea: consciousness might influence physical reality. It was as if the universe waited for an observer before deciding how to manifest itself.

Physicists like Max Planck, the father of quantum theory, embraced this implication with remarkable boldness. Planck once stated:

"I regard consciousness as fundamental. I regard matter as derivative from consciousness."

This declaration floored me. Planck, a scientist deeply rooted in empirical study, had arrived at a conclusion that echoed Plato's ancient idealism. The physical world, traditionally viewed as the foundation of reality, was now cast as secondary to consciousness, a profound reversal of materialist assumptions.

The implications of these findings are staggering. If consciousness is fundamental, what does this mean for the nature of reality? Could the universe itself be a construct of mathematical principles, shaped by a higher order of consciousness or thought? This idea finds resonance in the notion of the "holographic principle," a theory in physics suggesting that the three-dimensional universe we perceive may be a projection of information encoded on a two-dimensional surface. Such concepts blur the line between materialism and idealism, inviting speculation that reality might be both mathematical and mental, an intricate dance between numbers and mind.

Moreover, the integration of these ideas with advancements in neuroscience further challenges the boundaries between mind and matter. Modern brain imaging studies reveal the intimate link between neural activity and subjective experience, yet they also raise profound questions. Are our minds merely products of physical processes, or do they tap into a universal consciousness?

Perhaps the truth lies not in choosing one framework over the other, but in recognizing their interplay. Could the universe be a mathematical construct, a reflection of a deeper consciousness? Might the physical and the mental be two sides of the same coin, interwoven in ways we are only beginning to comprehend?

The deeper I delve into these questions, the more I realize how little we truly know. The universe, whether it is a material machine, a mathematical construct, or a manifestation of consciousness, remains an enigma. Yet, this uncertainty is what drives human curiosity, the relentless quest for understanding, unifying principles, and the ultimate truth about reality.

Gödel's Incompleteness Theorems

When I first encountered the work of Kurt Gödel, I was both enthralled and unsettled by its implications. Gödel's two Incompleteness Theorems struck at the very foundations of mathematics, suggesting limits to what even the most rigorous logical systems can achieve.

In the first theorem, Gödel demonstrated that in any sufficiently powerful formal system, capable of describing basic arithmetic, there exist statements that cannot be proven true or false using the rules of the system itself. These "undecidable" propositions lurk in the logical structure, immune to conclusive resolution from within. It was as if the mathematical framework, designed to capture all truths, contained shadowy corners that eluded its own methods of proof.

The second theorem went further, showing that such a formal system could never prove its own consistency (assuming it actually is consistent). This is the intellectual equivalent of trying to lift oneself off the ground by tugging on one's own hair; there is an inherent limitation to self-verification that no amount of brilliance can overcome.

Gödel's theorems have philosophical consequences that reach far beyond pure mathematics. They whisper a provocative suggestion: that reality, if viewed as a logical or mathematical structure, might contain truths that cannot be pinned down or fully comprehended from inside the framework of our physical universe. If consciousness is indeed fundamental, as idealists like Plato and Berkeley suggest, and as quantum theorists like Planck were bold enough to entertain, then Gödel's work hints at a realm of truths forever lurking beyond the edges of our cognitive apparatus. What if our very notion of "truth" is forever bound to the vantage point of the perceiver? Gödel's theorems crack open the door to the possibility that some realities remain irreducible mysteries, inaccessible not because we lack intelligence, but because the cosmos itself might be built on principles that defy complete internal explanation.

In this light, Gödel's findings stand as a testament to both the power and the humility of formal reasoning. They do not undermine the importance of science or mathematics; rather, they remind us that no matter how advanced our systems become, they may never capture every facet of the Real. The deeper we dig into the foundations of logic, the more we find logical bedrock giving way to vast, unfathomable abysses. It is in these abysses that new forms of understanding might arise, if we dare to imagine a broader perspective that transcends the finite boundary of formal systems.

Information Theory

Where Gödel's theorems reveal the inherent limits of formal systems, Claude Shannon's Information Theory opens a different yet equally provocative window into the nature of reality. Shannon's work, initially designed to optimize data transmission over noisy communication channels, soon evolved into a universal framework for understanding "information" in practically every domain, biology, physics, neurology, and even cosmology. According to Information Theory, the fundamental currency of the universe may not be matter or energy alone, but bits of information, the measure of uncertainty or entropy that gets resolved as a system evolves or as an observer learns something new.

This perspective is nothing short of revolutionary. If everything, from a photon's path in a double-slit experiment to the neural firing patterns that give rise to a thought, can be described in terms of information, then reality itself might be fundamentally informational. Combine this insight with the idealist notion that consciousness plays a central role in manifesting the physical world, and a striking speculation emerges: the physical universe could be a cosmic-scale information processor, a grand network of interacting bits that only take on tangible form when observed, measured, or decoded. Such an idea resonates with the holographic principle.

Philosophically, Information Theory stirs the pot of existential questions. If bits of data are the bedrock of everything we perceive, what is the role of meaning, purpose, or identity? Are we merely ephemeral patterns of information written briefly into the cosmic ledger, only to be erased or repurposed in an ongoing computational dance? Or does consciousness, as the ultimate decoder, bring something transcendent to these bits, infusing them with purpose and context? Bold theorists argue that consciousness might not just be a by-product of intricate data processing; perhaps it is an active architect, selecting among probabilities to shape what we call "reality."

Moreover, Information Theory intersects intriguingly with Gödel's Incompleteness: if certain truths within a logical structure remain forever hidden, might there be "information" in the universe that remains perpetually inaccessible? It is as though our cosmic puzzle includes pieces that can never be perfectly slotted into place, no matter how cleverly we compute or observe. At the same time, the staggering success of information-based technologies, ranging from quantum computing experiments to brain-computer interfaces, suggests that harnessing and interpreting data remains one of our most potent means of approaching the Real.

Thus, in this digital age, we find ourselves standing on a precipice. We see a world in which the immaterial, ideas, data, consciousness, seems to hold as much weight as the material, if not more. Gödel warns us of epistemic limits, while Shannon and the heirs of Information Theory beckon us to explore reality through the lens of bits and codes. Neither perspective offers the final word; both merely expand our horizons, forcing us to question whether our cherished notions of matter, space, and time are but reflections of a deeper, intangible information matrix. The stage is set for a new synthesis, one in which logic meets consciousness and information meets mystery, unveiling a grander, more elusive vision of reality than we ever dared imagine.

My Big TOE

Thomas Campbell, a physicist, consciousness researcher, and author, is renowned for his ground-breaking trilogy *My Big TOE* (Theory of Everything). This series presents a unified theory that seeks to explain the nature of reality, the purpose of existence, and the role of consciousness in shaping the universe. Campbell's work bridges the realms of science and spirituality, offering a logical framework that integrates quantum mechanics, metaphysics, and personal experience into a cohesive model of reality.

Campbell began his career as a physicist, collaborating with organizations such as NASA and focusing on modelling large, complex systems. His expertise in simulations and systems theory paved the way for his exploration into consciousness and reality's deeper mechanisms. In the 1970s, he worked with Robert Monroe, a pioneer in out-of-body experiences and founder of the Monroe Institute. This partnership exposed him to phenomena that traditional physics could not explain, prompting his quest to create a theory uniting subjective and objective realities.

At the core of Campbell's theory is the assertion that consciousness is the fundamental fabric of reality. Unlike conventional scientific views that treat consciousness as a by-product of physical processes, Campbell posits that the physical universe emerges from consciousness. His theory revolves around several key principles:

- **Reality as a Digital Simulation**

  Campbell suggests that our universe operates as a digital simulation orchestrated by a Larger Consciousness System (LCS), a cosmic "computer" designed to evolve consciousness. In this simulation, individuated units of consciousness (IUOCs) learn and grow through experiences.

- **Consciousness as the Prime Reality**

  Consciousness, not matter, is the true substance of existence. Physical space, time, and matter are emergent properties of consciousness. Our individual consciousness's are subsets of the LCS, participating in the simulation to foster growth and learning.

- **Entropy Reduction**

  The purpose of existence is to reduce entropy, or disorder, within the consciousness system. By choosing love and cooperation over fear and self-interest, consciousness evolves toward lower-entropy states, promoting harmony and effectiveness.

- **The Role of Free Will**

  Free will is integral to this simulation. Every choice, no matter how small, contributes to the evolution of consciousness. This dynamic learning environment allows for mistakes, experimentation, and growth.

- **Non-Physical Realities**

  The physical universe is just one of many virtual realities. Non-physical phenomena such as dreams and out-of-body experiences are also part of the larger consciousness system, each serving distinct purposes.

- **The Evolution of Consciousness**

  Consciousness's ultimate goal is growth. Ethical and constructive choices align with this purpose, while fear and selfishness hinder progress.

- **The Fear vs. Love Dichotomy**

  Fear corresponds to high entropy, leading to disorder, while love represents low entropy, fostering order and growth. Campbell emphasizes love as the driver of both personal and collective evolution.

- **Experiential Evidence**

  Campbell advocates for direct exploration of consciousness through meditation, lucid dreaming, and other practices, emphasizing that personal experience is crucial to understanding reality.

Brian Whitworth : the Virtual Construct Theory

Dr. Brian Whitworth took both perspectives seriously, acknowledging the empirical success of materialism while also grappling with the philosophical insights of idealism.

Brian Whitworth, a physicist and philosopher, proposed a complementary perspective on the nature of reality, suggesting that our universe might be a virtual construct. According to Whitworth, reality operates as a simulation running on information processing systems outside space-time. His theory addresses key limitations in traditional materialism and idealism by merging their strengths.

Materialism emphasizes empirical predictability and physical laws but struggles to explain phenomena like consciousness, quantum indeterminacy, and the fine-tuning of universal constants. Whitworth argued that materialism overlooks the possibility that the physical universe is a manifestation of something deeper.

Idealism posits that reality is fundamentally mental. Whitworth highlighted its alignment with quantum mechanics, particularly the observer-dependent nature of particles. He proposed that the universe's "pixels", atoms, particles, and forces, are the visible outputs of computational processes occurring beyond our perception.

Whitworth's model depicts reality as a virtual reality game:

- **Hardware:** Information processors exist outside space-time and are inaccessible to us.
- **Software:** Universal laws govern the behaviour of simulated reality.
- **User:** Conscious observers interact with the virtual construct, shaping their experiences within it.

This view aligns with Einstein's reflections on the illusory nature of time and space, as well as Max Planck's assertion that consciousness is fundamental.

Both Campbell and Whitworth's theories align with perspectives from digital physics and ancient philosophies. *Edward Fredkin's* cellular automaton model and Stephen Wolfram's computational universe hypothesis suggest that physical laws arise from simple rules iterating over time. Similarly, philosophers like Schopenhauer and traditions like Hinduism and Buddhism describe reality as an illusion or projection of deeper truths.

Furthermore, Donald Hoffman's theory that perception is a survival-oriented "user interface" rather than objective truth echoes the simulation concept. Additionally, Gödel's incompleteness theorems underscore the limitations of purely materialist descriptions of reality, suggesting that the universe's true nature lies beyond our current understanding.

Together, these ideas transform materialism's focus on physicality and idealism's emphasis on mind into a unified vision of an informational reality. They challenge us to rethink existence itself as a participatory simulation shaped by consciousness, evolving through the interplay of free will, love, and experiential growth.

The Simulation Argument

As I wrestled with the opposing perspectives of idealism and materialism, I stepped into the simulation argument. Building on the mentioned foundations, *Nick Bostrom* built his argument. Bostrom posited that if advanced civilizations could create highly realistic simulations, it is statistically likely that we live in one of them. This hypothesis bridges materialism and idealism: the simulation is a material construct, bits of data running on hardware, yet its reality depends on perception.

The simulation hypothesis can be viewed as a modern iteration of Plato's Allegory of the Cave, René Descartes' philosophical scepticism, George Berkeley's immaterialism, and Max Planck's quantum theory, reinterpreted through the lens of technology and computational advances. It suggests that age-old philosophical questions about existence and perception might now find new answers in the digital age, further blurring the boundaries between what is "real" and what is "constructed."

Bostrom's argument is not merely speculative; it is grounded in logic and probability, proposing that at least one of three propositions must hold true:

1. The probability that a species at our level of development avoids extinction before reaching a "post-human" stage is negligibly small.
2. Post-human civilizations are highly unlikely to run a significant number of ancestor-simulations.

3. We are almost certainly living in a computer simulation.

At the core of the simulation hypothesis is the principle of substrate independence. This philosophical stance posits that consciousness does not inherently depend on biological matter, such as a brain. Instead, mental states and subjective experiences could theoretically emerge from computational processes on any sufficiently advanced substrate, such as silicon-based processors.

This forms the foundation for the idea that a computer could replicate the human mind if it simulates the right computational patterns. Importantly, the argument does not require strong assumptions about artificial intelligence; it only necessitates that a detailed simulation of a human brain could produce conscious experiences.

For the simulation argument to hold, future civilizations must possess computational power far beyond what we can currently imagine. Bostrom estimates that even with modest advancements, post-human civilizations could have the capacity to simulate entire human histories, including detailed environments, on a massive scale.

Consider this:

- The human brain requires approximately $\mathbf{10^{14}}$ ***to*** $\mathbf{10^{17}}$operations per second to function. Simulating the entirety of human history (~100 billion humans) would require $\sim\mathbf{10^{33}}$***to*** $\mathbf{10^{36}}$operations.
- Theoretical models for post-human computers, such as planetary-mass computers, suggest they could perform $\mathbf{10^{42}}$ operations per second, vastly exceeding the requirements for running a single ancestor-simulation.

Even if these estimates are off by several orders of magnitude, the technological feasibility of ancestor-simulations remains plausible.

The argument becomes rigorous when analysed mathematically. Bostrom introduces the following notations:

- f_p**:** The fraction of civilizations that reach the post-human stage.
- **N:** The average number of ancestor-simulations run by a post-human civilization.
- **H:** The average number of individuals who have lived in a civilization before it reaches a post-human stage.
-

The fraction of observers with human-type experiences that are simulated is given by:

$$f_{sim} = \frac{f_p\, N\, H}{f_p\, N\, H + H}$$

- Let f_I be the fraction of post-human civilizations interested in running ancestor-simulations.
- Let N_I be the average number of ancestor-simulations run by these interested civilizations.

Then:

$$f_{sim} = \frac{f_p f_I N_I}{(f_p f_I N_I)+1}$$

Since N_I is likely to be extremely large, the equation simplifies to show that the fraction f_{sim} approaches 1 unless f_p or f_I is nearly zero.

1. **Extinction Before Post-humanity ($f_p \approx 0$)**: If almost no civilizations reach the post-human stage, it implies that intelligent life is either exceedingly rare or self-destructive. This aligns with the "Great Filter" hypothesis, suggesting that catastrophic events or self-inflicted technological disasters prevent civilizations from advancing.
2. **No Interest in Ancestor-Simulations ($f_1 \approx 0$)**: Advanced civilizations may converge toward ethical principles or practical constraints that deter them from running simulations. For example:
    - Simulations might be considered morally reprehensible due to the suffering inflicted on simulated beings.
    - Post-humans may lose the desire for such simulations, seeing them as trivial or ethically unnecessary.
3. **We Are in a Simulation ($f_{sim} \approx 1$)**: If both f_p and f_I are non-zero, the overwhelming probability is that we are living in a simulation. This follows from the sheer number of potential simulated observers compared to biological ones.

If we accept the simulation hypothesis, our understanding of reality shifts dramatically:

- The universe we observe is not "fundamental reality" but a construct within a higher-level reality.
- Simulated beings (us) may never have access to the underlying nature of their simulation or the motives of their creators.

The simulation argument also raises theological and ethical questions:

- Simulators, akin to gods, could intervene in ways that violate the physical laws of the simulated world.
- If simulators reward or punish behaviours, this could create a "naturalistic theogony" influencing morality.

The simulation hypothesis reconciles materialism and idealism by proposing that reality is both a computational construct and a perceptual phenomenon: materialism is reflected in the idea that the simulation exists as data and algorithms on a physical substrate, while idealism is embodied in the notion that reality, as perceived by simulated beings, is entirely dependent on their conscious experience.

The Cosmic Code

Understanding the universe is like deciphering a book written in an unknown language, where every page reveals more questions than answers.

I'm going to take a wild guess and assume you interact with the digital world every day. Maybe you use your phone for social media, emails, or messaging, or perhaps you use a computer to play games, check your bank account, or send messages to friends. But here's the question: does using these technologies mean you truly understand how they work?

Sure, you know how to open your phone, tap the Gmail icon to compose a new message, or move your mouse across the screen to open a folder. You understand how to interact with them on the surface. But do you know what happens underneath? Do you understand the intricate processes that make these actions possible?

Consider this piece of code:

```
print("Hello, World!")
```

Take a moment to look at it. Can you guess what it does or what its purpose might be?

This is a simple Python program. When you run it, the phrase *"Hello, World!"* will be displayed on your screen. You don't need a degree in computer science to figure out its purpose, it's straightforward and self-explanatory.

Now, look at this one. Can you guess its purpose?

```
int complex_recursive(int n, int (*fn)(int)) {

    return n == 0 ? 1 : OP(n) + FUNC(fn, n - 1);

}

int helper(int n) {

    return n < 2 ? n : complex_recursive(n, helper);

}

int main() {

    int input = 7;

    printf("Result: %d\n", complex_recursive(input, helper));

    return 0;

}
```

This C program is far from simple. It uses recursion, macros, function pointers, and bitwise operations to perform a convoluted computation. The complex_recursive function processes input through a series of transformations that include XOR, left shifts, and conditional logic. The use of a macro (FUNC) and a helper function adds layers of abstraction, making the code difficult to interpret at a glance. Even if you have a degree in computer science, unless you're familiar with C and the nuances of its syntax, understanding this code might be a challenge.

Similarly to how you use technology every day without fully understanding how it works, you also exist in this universe and experience reality daily. You breathe, drink water, touch, feel, but does simply existing mean you understand how your existence is possible? Does living in this reality mean you have enough knowledge to draw definitive conclusions about the nature of reality itself? Does it provide the irrefutable evidence needed to prove that this is the primary reality and not merely a reflection or a secondary reality?

Now, here's the important question: Can you judge the nature of reality if you don't understand the underlying principles of the universe? If you're unfamiliar with the "laws of reality", the fundamental rules that govern the cosmos, how can you accurately assess what reality is?

Think of the universe as a complex code like the C program above. Its laws are intricate, interconnected, and often counterintuitive. To even begin deciphering them, you need to understand the basics.

The arguments that our reality is not the primary reality or that we are living in a simulation are often labelled as conspiracy theories, fiction, or philosophical musings, dismissed as "la-la-land theories" by those who have little to no understanding of how the universe or reality works. What we're trying to do here is to go through the "codes" of the universe line by line and uncover the true nature of reality. Our goal is to show that the idea that our reality is not the primary reality is not fiction, but a reflection of the written laws of the universe.

We're going to look at the universe's "codes", the numbers, and see what they reveal about the nature of existence. Our conclusions will not be based on hypotheses or abstract philosophy but on the observable, written laws of the universe. We'll follow the numbers, the equations, and the patterns to their logical conclusions.

This book isn't about teaching you how to solve equations or perform advanced calculations. Instead, it's about understanding the essence of these equations, what they represent, why they matter, and how they help us make sense of the world.

To explore the nature of reality, we'll start with foundational concepts from physics. These ideas are deeply intertwined with mathematics, but don't worry, you won't need an advanced degree to follow along. Everything will be explained in an accessible and engaging way, ensuring that even the most complex ideas feel approachable.

Let's face it: trying to understand something without the right context can feel overwhelming, and boring. I've always found it frustrating when books or articles assume a level of knowledge without providing the necessary background. That's not the approach we'll take here.

We'll begin with some of the most significant discoveries in physics, Newton's Laws, Maxwell's Equations, Einstein's Relativity, and more. Together, we'll build a framework for understanding these concepts. The goal isn't to turn you into a physicist but to equip you with the tools and insights needed to appreciate the profound questions this book seeks to explore.

You don't need to understand every equation or the intricate details of how they work. Instead, focus on grasping the core concepts of each law and what they aim to explain. By grounding ourselves in these principles, we'll be better prepared to delve into the deeper mysteries of existence.

Newton's Laws of Motion: The Foundations of Classical Mechanics (1687)

In 1687, the world of science took a monumental leap forward with the publication of Sir Isaac Newton's *Philosophiæ Naturalis Principia* Mathematica, often simply called the *Principia*. Within its pages lay the foundational principles of classical mechanics and the law of universal gravitation, ideas that would shape our understanding of the physical universe for centuries to come. But to truly appreciate these ground-breaking concepts, we must journey back in time to uncover how Newton arrived at these ideas, what inspired them, and the genius behind the mathematics and reasoning.

Imagine a young Isaac Newton, a curious boy born in 1642 in the rural English village of Woolsthorpe. Growing up, Newton displayed an insatiable curiosity about the world around him. As he matured, he became particularly fascinated by the movements of celestial bodies and the forces of nature. It was during the Great Plague of 1665, when Cambridge University temporarily closed its doors, Newton, a student at the time, retreated to his family's estate and began what he later referred to as his *"annus mirabilis"* or "wonderful year."

One popular story from this period is the tale of the apple. While often romanticized, it is rooted in truth. Newton himself recalled that seeing an apple fall from a tree sparked his curiosity. He began to wonder why the apple fell straight down and not sideways or upward. What force compelled it toward the Earth? Could this same force be responsible for the motions of the Moon and planets?

At the heart of Newton's thinking lay a revolutionary idea: the same set of principles governs the motions of both earthly objects and celestial bodies. Until then, most people believed that the heavens and the Earth followed entirely different rules. Newton challenged this notion by proposing a universal law of gravitation, a single force that explained both the fall of an apple and the orbit of the Moon.

To build his theories, Newton relied on the shoulders of giants. He drew inspiration from earlier thinkers like Galileo Galilei, who studied the motion of objects, and Johannes Kepler, who described the elliptical orbits of planets around the Sun. Galileo's experiments with inclined planes demonstrated that objects accelerate uniformly under gravity, while Kepler's laws of planetary motion revealed the mathematical patterns behind celestial movements. Newton sought to unify these observations under a single framework.

The journey to Newton's laws of motion began with careful observation and experimentation. The first law, often called the law of inertia, states:

An object will remain at rest or in uniform motion in a straight line unless acted upon by an external force.

This principle built upon Galileo's concept of inertia, emphasizing that motion does not require a continuous application of force but rather persists unless disrupted.

The mathematical expression for this is simple yet profound:

$$\sum F = 0 \rightarrow \textit{constant velocity}$$

Where $\sum F$[10] is the net force acting on an object. If the net force is zero, the object remains in its current state of motion, either at rest or moving in a straight line at constant speed. This idea challenged Aristotelian physics, which held that objects naturally come to rest unless acted upon by an external agent, thereby setting the stage for a more accurate understanding of motion.

This principle built on Galileo's insights. Imagine pushing a toy car across a smooth floor. It might eventually come to a stop, but that's because of friction, an external force. If there were no friction, the car would keep moving forever. Newton realized that motion itself doesn't require a force; rather, a force is needed to change motion. This idea shattered the ancient belief that a continuous force was required to keep an object moving

In a simulation context, inertia could be interpreted as a baseline parameter programmed to maintain stability in the simulated environment. Just as virtual objects in video games remain stationary until acted upon by external inputs (e.g., forces applied by the player or the game's physics engine), real-world objects obey this rule universally.

The idea of uniform motion in a straight line is particularly interesting, as it implies that motion itself is not inherently "active" but rather a default state requiring no additional input to sustain. This echoes principles in computational efficiency, where systems default to simple, resource-conserving states unless acted upon. The need for an external force to alter motion could be seen as a safeguard against chaotic, uncontrolled behaviour, ensuring the simulation remains orderly and predictable.

The second law of motion provides a mathematical framework to understand how forces affect motion:

The acceleration of an object is directly proportional to the net force acting on it and inversely proportional to its mass.

Mathematically, this is expressed as: $F = ma$. Where F is the force, m if the mass of the object, and a is its acceleration. Suppose you're pushing a shopping cart. If you push harder (increase the force), the cart accelerates faster. If the cart is heavier (greater mass), it takes more effort to achieve the same acceleration. This simple yet profound equation ties together force, mass, and motion in a way that is both intuitive and precise.

From a simulation perspective, this law could be interpreted as a programmed rule for managing interactions within the system. The proportionality constant mmm (mass) could be seen as a "variable" assigned to each object, determining how it reacts to applied forces.

What's particularly notable is the scalability of this equation. Whether applied to subatomic particles or massive celestial bodies, the relationship holds. Such scalability suggests an underlying computational simplicity, an elegant algorithm capable of governing interactions at all scales without the need for separate rules. This uniformity aligns with principles of efficient code design, where a single function or formula governs multiple scenarios.

Moreover, the second law introduces a time-dependent element: a (acceleration) represents a change in velocity over time. The existence of time as a measurable, continuous parameter further supports the idea of a structured system with a "clock" or temporal framework built

[10] Σ (Sigma) is a Greek letter pronounced as "sig-ma." In mathematics, it represents the summation or total of a set of values.

into its code. Without such a framework, the simulation's events could not unfold in a coherent, sequential manner.

The third law of motion is famously succinct:

For every action, there is an equal and opposite reaction.

Picture a swimmer pushing against the wall of a pool. As they push backward, the wall pushes them forward with equal force. This law isn't just about obvious interactions; it's the foundation for phenomena as diverse as rocket propulsion and walking on the ground. Every force comes in pairs, and these pairs act on different objects.

Newton's discoveries introduced the concept of a *clockwork universe*, a metaphor for a cosmos functioning with precision and predictability, much like a finely tuned machine. By unifying celestial and terrestrial mechanics under universal laws, Newton demonstrated that the same mathematical principles governed the orbits of planets, the fall of an apple, and the motion of tides. This perspective suggested that the universe operated according to immutable laws, which, once understood, could predict every event in nature with certainty. This deterministic view resonated deeply with Enlightenment thinkers, inspiring the belief that the universe was akin to a giant clock set into motion by an initial creator, and then left to run autonomously. In such a system, every action and reaction could, in theory, be calculated and anticipated, painting a picture of an orderly, predictable cosmos where randomness played no role, a concept that would shape science and philosophy for centuries.

In a simulation, this law could be interpreted as a balancing mechanism designed to maintain equilibrium. Without such a rule, interactions within the system could lead to net increases or decreases in energy, causing instability over time.

Interestingly, the symmetry of action and reaction suggests an underlying logic of fairness or consistency, akin to how simulations enforce rules to avoid glitches or unexpected behaviour. The universality of this law, even at scales where classical mechanics begins to break down, hints at its fundamental importance in maintaining the coherence of the simulated environment.

Universal Gravitation

Armed with these three laws, Newton turned his attention to the heavens. He realized that the force keeping planets in orbit around the Sun was the same force that pulled an apple to the ground: gravity. But how could he quantify this force?
Newton proposed his law of universal gravitation:

Every particle of matter in the universe attracts every other particle with a force that is proportional to the product of their masses and inversely proportional to the square of the distance between them.

His gravitational formula,

$$F = G\,\frac{m_1 m_2}{r^2}.$$

Where G is the gravitational constant, m_1 and m_2 are the masses of two objects, and r is the distance between their centres, demonstrated the universality of natural laws. This equation elegantly explains why the gravitational pull weakens with distance but grows stronger with larger masses. For example, the Earth's gravity is stronger than the Moon's because the Earth has more mass. Similarly, the gravitational pull weakens as astronauts travel farther from Earth. Newton's laws painted a picture of a clockwork universe, a system where every motion and interaction could, in principle, be predicted with mathematical precision.

At its heart lies the gravitational constant G, a fixed value that seems to be finely tuned to allow the formation of stars, planets, and life as we know it. But why does G have the precise value it does?

In the context of a simulation, this law, and particularly the inclusion of G, could be interpreted as a governing parameter set by the simulation's designers. The inverse-square relationship, where the force diminishes with the square of the distance, is strikingly efficient: it allows gravitational interactions to extend infinitely but with diminishing impact, avoiding runaway accumulation of forces at large scales while ensuring stability in local systems. This mathematical precision mirrors the kind of optimization seen in programming, where algorithms are crafted to balance complexity and computational efficiency.

Furthermore, the presence of G suggests a rule deliberately embedded to regulate the strength of gravity across the entire "box" of the universe. A slight deviation in G's value could lead to catastrophic instability, too high, and the universe would collapse under its own weight; too low, and galaxies and stars could never form. This fine-tuning is reminiscent of parameters in simulations, where constraints are calibrated to achieve desired outcomes.

The inverse-square law itself could also be seen as an artefact of a three-dimensional spatial framework within the simulation. In higher or lower-dimensional universes, the relationship between force and distance would differ. This consistency suggests that our universe operates under specific dimensional parameters, potentially programmed as part of the simulation's design. Newton's law of gravitation, then, may not merely describe a natural phenomenon but act as evidence of the calculated constraints underlying our existence.

Newton's laws collectively outline a universe that behaves with machine-like precision and predictability, attributes that align closely with the idea of a simulation. The laws of motion, with their scalable simplicity, time-dependence, and symmetry, suggest a rule-bound system designed for stability and coherence. Taken together with the law of gravitation, they paint a picture of a universe that operates not as a chaotic, self-emergent system but as one that could be purposefully engineered, governed by deterministic principles akin to the parameters of a simulation.

Newton's ground-breaking discoveries laid the foundation for a mechanistic view of the universe, a vast, orderly clockwork governed by mathematical laws. His laws of motion and universal gravitation appeared to reveal an elegant simplicity: every interaction, from the fall of an apple to the orbit of planets, was dictated by predictable forces.

Yet, this very predictability invites an intriguing hypothesis: could such a deterministic framework itself be evidence that the universe operates as a simulation? In a simulation, a defined set of rules, akin to Newton's laws, would guide the behaviour of all objects and interactions, just as programming code governs the mechanics of a virtual world.

The idea of a universe in a box resonates here; Newton's equations, as well as their successors in quantum mechanics and general relativity, suggest that all phenomena are constrained by a fixed set of principles. These principles, while seemingly infinite in their complexity, could themselves be the product of a designed system, a cosmic "code" that dictates the rules of existence within the confines of the simulated universe.

Pierre-Simon Laplace: Laplace's Demon (1814)

One of the most fascinating intellectual leaps in the history of philosophy and science was made by Pierre-Simon Laplace, an 18th-century French mathematician, physicist, and astronomer. Laplace's contributions to probability theory and celestial mechanics were ground-breaking, but it is his thought experiment known as "Laplace's Demon" that stands out as a cornerstone in the discourse on determinism, predictability, and the nature of reality itself.

Laplace, building upon the mathematical principles of Sir Isaac Newton, proposed a universe governed entirely by deterministic laws. Newton's laws of motion and universal gravitation had already demonstrated that the movements of celestial bodies could be precisely predicted given their initial positions and velocities. Inspired by this, Laplace extended the idea to every particle in the universe. He hypothesized the existence of an intellect, later referred to as "Laplace's Demon," which could know the precise location and momentum of every atom in the cosmos at a single moment in time. With this information, coupled with the laws of physics, such a being could calculate the entire past and future of the universe with absolute certainty.

In Laplace's words:

> *"We may regard the present state of the universe as the effect of its past and the cause of its future. An intellect which at a certain moment would know all forces that set nature in motion, and all positions of all items of which nature is composed, if this intellect were also vast enough to submit these data to analysis, it would embrace in a single formula the movements of the greatest bodies of the universe and those of the tiniest atom; for such an intellect nothing would be uncertain and the future just like the past would be present before its eyes."*

This profound idea was conceived long before the advent of modern computers or even the faintest notion of digital simulation. Laplace relied purely on mathematics and the mechanistic worldview of his time, extending the logic of Newtonian physics to its ultimate philosophical conclusion. Laplace's Demon is deeply rooted in the mathematics of his era. Newton had already laid the groundwork with calculus and the laws of motion, demonstrating how the forces acting upon objects determine their future trajectories. Laplace, a pioneer of probability theory, saw a deeper connection between predictability and the mathematical description of natural phenomena.

While probability often dealt with uncertainty, Laplace's deterministic view framed uncertainty as a limitation of human knowledge rather than a fundamental property of the universe. He developed mathematical techniques to study celestial mechanics, accurately predicting the movements of planets, moons, and comets. These successes reinforced his belief that the universe operated like a grand, intricate machine, one that could theoretically be understood in its entirety given sufficient information and analytical power. The "demon" became a metaphor for ultimate knowledge, a being capable of performing calculations on a scale that transcended human capabilities.

Though Laplace never used the term "simulation," his thought experiment resonates strongly with modern ideas of computational models and simulated realities. In essence, Laplace's Demon envisions the universe as a calculable system, much like a computer program where the state of every element determines the next. This deterministic framework aligns intriguingly with the concept of a simulated universe in a box, where all variables are predefined, and outcomes are generated through mathematical rules.

Laplace's Demon also highlights the philosophical implications of living in such a simulation. If reality is deterministic and calculable, then every moment of existence, past, present, and future, is already encoded in the initial conditions and the governing laws. For those exploring the notion of a simulated universe, Laplace's thought experiment provides a historical lens through which to examine how human minds have grappled with the idea of predetermination and the boundaries of knowledge.

While Laplace's Demon is an elegant representation of determinism, it has faced significant challenges over time. The emergence of quantum mechanics in the 20th century introduced the idea that certain events at the atomic and subatomic levels are inherently probabilistic. This undermines the classical deterministic worldview that Laplace envisioned.

Chaos theory further complicates the picture by showing how small changes in initial conditions can lead to vastly different outcomes, making long-term predictions exceedingly difficult even in deterministic systems. These challenges do not completely debunk the idea of a simulated universe. Instead, they raise an intriguing possibility: what if the apparent randomness and chaos in our universe are intentional features of the simulation? They could serve to make the simulation appear more "organic" or unpredictable to its inhabitants. After all, a simulation that is too rigidly deterministic might seem artificial to its conscious entities, us.

Maxwell's Equations of Electromagnetism (1861)

The story of Maxwell's equations is a tale of curiosity, ingenuity, and the relentless pursuit of understanding nature's most fundamental forces. To appreciate their significance, we must journey through time, following the breadcrumbs of discoveries that led to their creation. These equations describe how electric and magnetic fields interact and propagate, forming the foundation of classical electromagnetism and much of modern physics.
Ancient civilizations were intrigued by peculiar phenomena. Greek philosophers noticed that rubbing amber with cloth could attract small objects like feathers. They named this strange force "*elektron*," the Greek word for amber. They also discovered lodestones, naturally magnetized rocks that could attract iron. These observations laid the groundwork for our understanding of electricity and magnetism, though the connection between the two was far from understood.

In the 18th century, Benjamin Franklin experimented with lightning, proving that it was a form of electrical discharge. He introduced the concept of positive and negative charges. Around the same time, Charles-Augustin de Coulomb formulated the law describing the force between electric charges, similar to Isaac Newton's law of gravitation.

Hans Christian *Ørsted*, a Danish physicist, stumbled upon the link between electricity and magnetism in 1820. During a lecture, Ørsted noticed that a compass needle deflected when placed near a current-carrying wire. This revelation was ground-breaking: electricity could create magnetism. Ørsted's observation spurred a wave of discoveries. André-Marie Ampère derived mathematical laws describing the magnetic fields produced by electric currents. Michael Faraday in England and Joseph Henry in the United States independently discovered electromagnetic induction: a changing magnetic field could induce an electric current. Faraday's experiments led to the concept of the electromagnetic field, an invisible "web" connecting electricity and magnetism. However, while Faraday's insights were profound, his explanations were largely qualitative.

James Clerk Maxwell, born in Scotland in 1831, was a child prodigy with a passion for mathematics and physics. Captivated by Faraday's work, Maxwell realized that rigorous mathematics was needed to fully describe these ideas.
Maxwell began by unifying the various discoveries of his predecessors. He started with four key observations:

I. *Electric charges create electric fields (Coulomb's law).*
II. *Electric currents and changing electric fields produce magnetic fields (Ørsted's and Ampère's work).*
III. *Changing magnetic fields induce electric fields (Faraday's law of induction).*
IV. *Magnetic monopoles (isolated north or south magnetic poles) do not exist; magnetic fields are always dipolar.*

To capture these principles, Maxwell wrote down a set of equations, initially twenty in number, that mathematically described how electric and magnetic fields interact. He later refined these into the four elegant equations we know today.

Gauss's Law for Electricity

$$\nabla . E = \frac{p}{\varepsilon_0}$$

This equation describes how electric fields **(E)** are generated by electric charges. The symbol **∇** [11]. ***E*** (called the divergence of the electric field) measures how much the electric field spreads out from a point. On the right side of the equation, **ρ** is the electric charge density, representing how much electric charge is present in a given volume. The constant $\boldsymbol{\varepsilon_0}$[12] is the permittivity of free space, which quantifies how electric fields interact with the vacuum. This law essentially states that electric fields originate from positive charges and terminate at negative charges, with the total "spread" of the field proportional to the amount of charge present.

Imagine you're blowing air into a balloon. The air you blow represents the electric field **(E)**, and the balloon's surface represents the boundary where the field exits. The amount of air (electric charge, **ρ**) inside the balloon determines how much the balloon stretches (how much electric field spreads out). The material of the balloon (vacuum permittivity, $\boldsymbol{\varepsilon_0}$) determines how easily the air stretches it. Electric fields "leak out" from charges, just like air fills a balloon. Positive charges emit fields, and negative charges attract them.

---

[11] The symbol ∇ is called "nabla," which refers to its name as a mathematical symbol. However, it is often pronounced "del" when describing its use in operations such as gradient (∇E), divergence (∇·E), or curl (∇×E). Both terms are correct, with "nabla" emphasizing the symbol itself and "del" focusing on its function.

[12] ε0: Pronounced as "epsilon naught" or "epsilon zero," the permittivity of free space.

In a simulation context, this law could reflect how the system encodes local properties (charge density, ρ) into the larger, observable environment (electric field, E). The presence of ε_0 (the permittivity of free space) is particularly striking, as it acts as a universal constant regulating how electric fields behave. This constant might be seen as a "tuneable parameter" in the simulation, controlling the strength and spread of electric interactions. Its precise value ensures that atoms, molecules, and even macroscopic objects maintain stable configurations. A slight change in ε_0 could destabilize this structure, much like altering a key variable in a program could crash a simulation.

Additionally, Gauss's law elegantly demonstrates that electric fields always originate and terminate at charges. This rule enforces a level of predictability and symmetry within the system, suggesting a programmed consistency, much like rendering rules in a virtual environment.

Gauss's Law for Magnetism

$$\nabla \cdot B = 0$$

This equation explains a key property of magnetic fields **(B)**: they have no sources or sinks. The divergence $\nabla \cdot B$ measures the "spread" of the magnetic field, and this equation tells us that the spread is always zero. In simpler terms, magnetic field lines always form closed loops, and there are no isolated magnetic charges (magnetic monopoles). This contrasts with electric fields, which can begin and end on charges.

Think of magnetic field lines as loops of rope in your hands. No matter how you twist or stretch the rope, you can't find a loose end, it's always a loop. Unlike electric fields that have starting and ending points (charges), magnetic fields don't have "loose ends." Magnetic fields form closed loops, just like a rope with no beginning or end. There are no magnetic monopoles (single magnetic charges) to create isolated magnetic fields.

The lack of magnetic monopoles is an unsolved mystery in physics, but in a simulation, it could be explained as a design decision. The absence of monopoles simplifies the computational model of the universe, reducing the complexity of interactions between particles and fields. The closed-loop nature of magnetic fields also ensures conservation and coherence, preventing runaway effects that could destabilize the system. In programming terms, this constraint could function as a "debugging" mechanism, ensuring that every magnetic field interaction remains consistent with the rules of the simulation. The fact that theories like string theory predict monopoles, yet none have been observed, could hint that we are observing a curated system designed to exclude them.

Faraday's Law of Induction

$$\nabla \cdot E = \frac{\partial B}{\partial t}$$

This equation shows how a changing magnetic field **(B)** generates an electric field **(E)**. Here, $\nabla \times E$ (called the curl of the electric field) measures how the electric field circulates around a region. The term $\frac{\partial B}{\partial t}$[13] represents the rate at which the magnetic field changes with time, with the negative sign indicating that the induced electric field opposes the change in the magnetic field (a concept known as Lenz's Law). This principle underlies technologies like electric generators, where a changing magnetic field produces an electric current.

Picture a pot of water on a stove. Stirring the water (changing the magnetic field, **B**) creates a swirling motion (the electric field, **E**) that spirals outward. The faster you stir (change the magnetic field), the stronger the swirl (induced electric field). The negative sign tells us that the swirl resists the stirring motion, just as water resists the spoon. A changing magnetic field "stirs up" an electric field, creating motion and energy transfer, just as a swirling spoon does in water. This is how electric generators work!

In a simulated environment, this principle could be interpreted as an efficient algorithm for conserving and redistributing energy across the system. The negative sign in the equation, representing Lenz's law, ensures that induced electric fields always oppose the change in the magnetic field. This "negative feedback loop" mirrors control mechanisms in computational systems, where such loops are used to maintain stability and prevent runaway processes. Faraday's law could thus be seen as a rule built into the simulation to ensure energy transitions occur smoothly and predictably, avoiding chaos.

Ampere's Law with Maxwell's Correction

$$\nabla \times B = \mu_0 J + \mu_0 \varepsilon_0 \frac{\partial E}{\partial t}$$

This equation reveals how magnetic fields **(B)** are generated. The curl **∇×B** measures how the magnetic field curls around a region. On the right side, is the permeability of free space, a constant that describes how magnetic fields interact with the vacuum.

The term **(J)** represents the electric current density, which describes the flow of electric charges, while $\varepsilon_0 \frac{\partial E}{\partial t}$ is Maxwell's correction, accounting for the contribution of a changing electric field ($\frac{\partial E}{\partial t}$) to the magnetic field. This equation not only describes magnetic fields generated by electric currents but also explains how oscillating electric and magnetic fields sustain each other, enabling the propagation of electromagnetic waves, such as light.

Imagine a crowd in a stadium doing the wave. The wave travels around the stadium (magnetic field, **B**), but it needs something to start it: either people physically standing up and sitting down (electric current, **J**) or the energy from a prior wave (changing electric field, $\frac{\partial E}{\partial t}$). The stiffness of the wave depends on the crowd's energy (constants μ[14]$_0$[15] ***and*** ε_0).

---

[13] The symbol ∂ (called a "partial derivative symbol") is pronounced as "partial." It is specifically used in calculus to denote a derivative with respect to one variable while other variables are held constant, as often seen in multivariable calculus.

[14] **μ**: Pronounced as "mu," the Greek letter often representing permeability.

[15] **μo**: Pronounced as "mu naught" or "mu sub-zero," representing the permeability of free space.

A magnetic field can be created by moving charges (electric currents) or by a changing electric field. This dual mechanism explains how electromagnetic waves propagate.

The Mathematics of Waves: The Dance of Electric and Magnetic Fields

Maxwell discovered something extraordinary when analysing his equations. By combining Faraday's law and the modified *Ampère's law* (with Maxwell's correction), he derived a wave equation that describes how electric fields propagate through space:

$$\frac{\partial^2 E}{\partial t^2} = c^2 \nabla^2 E$$

A similar equation applies to magnetic fields. In this equation, **(C)** represents the speed of the wave, which Maxwell determined to be:

$$C = \frac{1}{\sqrt{\mu_0 \varepsilon_0}}$$

This result showed that the speed of electromagnetic waves is equal to the speed of light, a ground-breaking realization. Maxwell's work unified the phenomena of light and electromagnetism, demonstrating that light is, in fact, an electromagnetic wave. This insight laid the foundation for modern physics, connecting optics, electromagnetism, and wave theory into a single framework.

Imagine electric and magnetic fields as two dancers holding hands in a ballroom. When one moves (a changing electric field), it pulls the other along (inducing a magnetic field), and vice versa. This mutual pull allows them to move through the room in harmony, creating a continuous wave. Maxwell’s equations describe this beautiful dance, showing how light and all electromagnetic waves travel through space.

In this equation, μ_0 is the permeability of free space and ε_0 is the permittivity of free space. What’s striking about this equation is its suggestion that the speed of light is not just a natural phenomenon but a fundamental constant, hardcoded into the fabric of reality. This raises an intriguing question: why is the speed of light precisely what it is?

In a simulated environment, such constants could be viewed as parameters deliberately set by the creators of the simulation to ensure specific outcomes. The speed of light acts as a universal "speed limit," shaping the structure of causality and ensuring that no information can propagate faster than this set threshold. Such a constraint is analogous to frame rate or bandwidth limitations in a computer program, which dictate the flow of information to maintain coherence within the system.

The constant μ_0 and ε_0 too, appear to be intrinsic properties of the simulation’s "medium," ensuring that electromagnetic waves behave predictably.

The existence of electromagnetic waves as self-sustaining oscillations of electric and magnetic fields reveals a beautifully optimized mechanism for energy transfer. Maxwell's equations describe these waves with a symmetry and balance that suggest an engineered precision, where electric and magnetic fields sustain one another in perfect harmony. This interplay resembles feedback loops in computational systems, where predefined rules ensure coherence and stability.

Moreover, the relationship between the universal constants μ_0 , ε_0 and ccc highlights the elegance of a design optimized for stability and functionality. Much like how a programmer defines universal constants to maintain order in a simulation, Maxwell's framework encodes light as a wave of oscillating fields propagating at a precise universal speed. This symmetry and efficiency could be seen not merely as a description of physical phenomena but as evidence of a system intentionally crafted to reflect stability and coherence, akin to the mechanics of a deliberately engineered simulation.

Maxwell's equations go beyond describing physical phenomena; they reveal a universe governed by symmetry, balance, and optimization. From the absence of magnetic monopoles to the propagation of electromagnetic waves, these laws suggest a meticulously designed system, where constants like μ_0 , ε_0, and **c** act as parameters fine-tuned to ensure stability and coherence. Each equation, with its specific constraints and interactions, resembles the kind of rules one would encode into a simulation to produce a stable, predictable, and beautifully interwoven reality.

Max Planck's quantum theory (1900)

In the waning days of the 19th century, the world of physics found itself at a peculiar crossroads. The mechanistic universe that Sir Isaac Newton had described over two centuries earlier had served as a steadfast foundation for understanding nature. Yet, cracks were beginning to appear in this seemingly indomitable framework. One of these cracks came from an area that seemed innocuous at first: the study of light and heat.

The story of Max Planck and the birth of quantum theory begins here, in a field called thermodynamics, specifically the study of blackbody radiation. A blackbody, despite its name, is not an object of impenetrable darkness but rather an idealized material that absorbs all radiation falling upon it without reflecting any. Crucially, it also emits radiation when heated, and scientists in the late 19th century were puzzled by how to describe the distribution of this emitted energy across different wavelengths.

Max Planck, a German physicist with an unassuming demeanour but a tenacious mind, was born in 1858. By the 1890s, Planck was deeply immersed in the field of thermodynamics and had a growing interest in understanding the mysterious behaviour of blackbody radiation. At the time, scientists used classical physics to try to predict how energy from a blackbody would be distributed across various wavelengths. These efforts led to what became known as the "ultraviolet catastrophe."

Classical theories predicted that the energy emitted at shorter wavelengths, toward the ultraviolet end of the spectrum, would grow without bound. This was absurd; no blackbody in reality behaves this way. Experiments showed that energy peaks at a certain wavelength and then diminishes. The discrepancy between theory and experiment demanded an explanation.

Planck began his journey by attempting to reconcile these observations with the classical laws of physics. He was not initially seeking to overturn these laws but rather to find a mathematical model that fit the data. In October 1900, Planck presented a formula that accurately described the experimental results of blackbody radiation. But Planck's equation alone was not enough; he sought a deeper understanding of why it worked.

Planck's breakthrough came in December 1900, during a lecture to the German Physical Society. In an effort to derive his formula from first principles, he made a bold and counterintuitive assumption. He proposed that energy was not emitted continuously but in discrete packets, which he called "quanta." Each quantum had an energy proportional to the frequency of the radiation, with the constant of proportionality being a new fundamental constant, now known as Planck's constant **(h).** The relationship was expressed as:

$$E = hv$$

Here, **(E)** is the energy of a quantum, **(v)** (pronounced "nu") is the frequency of the radiation, and **(h)** is Planck's constant, an extraordinarily tiny number (~6.626 × 10^{-34} joule-seconds).

Imagine you are filling a jar with coins. Classical physics assumed that you could pour money into the jar like a fluid, smoothly and continuously. Planck's radical idea was akin to saying that money could only be added in fixed denominations, like pennies, nickels, or dimes. Similarly, energy, he proposed, could only exist in discrete amounts, not as a smooth continuum.

Just as pixels define the smallest unit of resolution in a computer screen, **h** defines the "resolution" of energy in the universe. The existence of such a precise, non-zero value suggests a universe that is quantized at its core, much like digital simulations that operate with discrete units of information. This granular structure may indicate that our universe is not continuous but instead operates on a set of programmed rules, where even energy comes in indivisible increments.

Planck himself was uneasy about the implications of his theory. He viewed quanta as a mere mathematical tool, a convenient fiction to solve a thorny problem. But the genie was out of the bottle. Planck's hypothesis laid the groundwork for quantum mechanics, a new branch of physics that would profoundly reshape our understanding of the universe.
Mathematically, Planck's law described the spectral energy density of blackbody radiation as:

$$E(v,T) = \frac{8\pi h v^3}{v^3} \cdot \frac{1}{e^{\frac{hv}{k_B T}} - 1}$$

In Planck's law, **E(v,T)** represents the spectral energy density, which is the amount of energy emitted per unit frequency per unit volume at a given frequency **(v)** and temperature **(T)**. The symbol **(h)** stands for Planck's constant, a fundamental constant of nature that relates energy and frequency, with a value of approximately 6.626 × 10^{-34} joule-seconds. The frequency **(v)** refers to the number of oscillations of the electromagnetic wave per second. The constant **(c)** denotes the speed of light in a vacuum, approximately 3.00×10^{8} meters per second, and it helps relate the frequency and wavelength of electromagnetic radiation. Boltzmann's constant, **(k_B)**, has a value of approximately 1.381×10^{-23} joules per kelvin and connects the macroscopic properties of temperature to the microscopic energy scales of particles. Finally, **(T)** is the absolute temperature in kelvin, which measures the thermal energy of the blackbody, influencing the intensity and distribution of radiation across different frequencies. Together, these constants and variables explain the energy distribution of blackbody radiation across all wavelengths and temperatures.

This equation incorporates fundamental constants: Planck's constant, the speed of light, and Boltzmann's constant. It captures the behaviour of blackbody radiation across all wavelengths and temperatures, resolving the ultraviolet catastrophe.

Classical physics struggled to explain the behaviour of blackbody radiation, leading to the infamous "ultraviolet catastrophe," where predictions allowed for infinite energy densities at high frequencies. Max Planck resolved this dilemma by introducing the concept of quantized energy, which placed natural limits on energy emission. This idea not only aligned with experimental data but also prevented the theoretical instability that would arise from infinite values. In a simulated framework, this quantization could be likened to an optimization mechanism, where energy constraints prevent computational overload, much like how bounded resource usage avoids crashing a computer program. Planck's solution thus mirrors the boundaries and constraints employed in simulations to maintain stability and functionality.

The quantization of energy fundamentally challenges the classical notion of a smooth, continuous universe, instead proposing a reality built on discrete, rule-based interactions, a characteristic commonly associated with digital simulations. By revealing that energy exists only in specific amounts, Planck's theory suggests a universe operating more like a digital program than a continuous analogue process. This notion extends to other quantized phenomena, such as angular momentum and atomic energy levels, reinforcing the view of a universe governed by fixed increments, reminiscent of binary logic in computational systems.

Planck's quantum theory also introduced a probabilistic interpretation of nature, later expanded upon in quantum mechanics. In a simulated environment, such probabilistic behaviour could represent a deliberate programming choice to introduce variability and complexity. By incorporating controlled randomness, the system avoids the predictability of purely deterministic processes, enabling emergent behaviours and a more dynamic, adaptable environment. This is akin to the use of randomization algorithms in simulations or video games, where unpredictability enhances realism and depth. The probabilistic nature of quantum mechanics, therefore, might reflect an intentional design feature, making the universe not only robust but also versatile and rich in possibilities.

Planck's work did not immediately overturn the classical worldview. Even he considered it an isolated correction. Yet, within a few years, Albert Einstein would seize upon the quantum idea to explain the photoelectric effect, another puzzle classical physics could not solve. Einstein's insights elevated Planck's quanta from a mathematical convenience to a cornerstone of reality.

Planck's quantum theory was the spark that ignited a revolution. It dismantled the deterministic clockwork universe envisioned by Newton and introduced a world governed by probabilities, uncertainties, and the strange behaviour of particles on the smallest scales.

For Planck, this journey was both thrilling and humbling. As he later reflected:

> *"A new scientific truth does not triumph by convincing its opponents but because a new generation grows up that is familiar with it."*

Thus, the quantum revolution began, not with a thunderous proclamation but with the quiet resolve of a physicist who dared to question the continuity of nature. Planck's quantum theory highlights a universe that operates not as a smooth, continuous system but as a quantized, rule-bound framework. The introduction of discrete energy packets, the resolution of the ultraviolet catastrophe, and the fine-tuned interplay of fundamental constants all point to a system with precise constraints and boundaries. These features are consistent with the characteristics of a **BOX**, a simulated box, where parameters are set, boundaries enforced, and algorithms optimized for stability and functionality. Far from being a random or chaotic cosmos, the universe revealed by Planck's work appears to be one governed by digital-like precision, offering compelling evidence for the universe in a box.

Special Relativity (1905)

In the early 20th century, the world of physics was teetering on the edge of a monumental transformation. Classical mechanics, the elegant framework constructed by Isaac Newton, had reigned supreme for over two centuries. It described the motion of planets, the fall of apples, and the mechanics of machines with unparalleled precision. But as humanity's understanding of the universe deepened, cracks began to appear in Newton's grand edifice. It was Albert Einstein, a young, unconventional thinker, who would lead the charge into a new era of physics. His work on relativity would not just rewrite the rules of physics but fundamentally alter humanity's understanding of space, time, and the cosmos.

The Birth of Special Relativity

In 1905, a 26-year-old Einstein published a series of ground-breaking papers in *Annalen der Physik*. Among these was "On the Electrodynamics of Moving Bodies," which introduced the Special Theory of Relativity. At the heart of the puzzle was light. James Clerk Maxwell's unification theory of electricity and magnetism raised a thorny question: if light is a wave, what medium does it travel through?

Scientists postulated the existence of a mysterious *"luminiferous aether,"* an invisible substance that filled the universe and acted as the medium for light waves. But experiments, most famously the Michelson-Morley experiment in 1887, failed to detect this aether.

Einstein approached the problem from a different angle. He began by accepting two seemingly incompatible principles:

I. *The laws of physics are the same in all inertial reference frames (frames of motion that are not accelerating).*

II. *The speed of light is constant, regardless of the motion of the source or the observer.*

These principles seemed to defy common sense. If you're on a train moving at 50 km/h and you throw a ball forward at 10 km/h, an observer on the ground would see the ball moving at 60 km/h. shouldn't light behave the same way? Yet experiments showed that the speed of light was always the same: about 300,000 kilometres per second.

Einstein realized that the problem wasn't with light but with the nature of space and time themselves. He proposed that time and space were not absolute but relative, they could stretch, shrink, and warp depending on the motion of the observer. This idea led to profound consequences:

Time Dilation

Time dilation describes how time slows down for someone moving very quickly compared to someone who is stationary. The equation,

$$t' = \frac{t}{\sqrt{1 - \frac{v^2}{c^2}}}$$

Means that the time experienced by a moving observer **(t')** [16]is less than the time experienced by someone stationary **(t).** The closer the speed of the moving observer (v) gets to the speed of light (c) the more time slows down for them. This is why astronauts traveling at near-light speeds in science fiction stories age more slowly than people on Earth. This isn't just science fiction; it's been confirmed with precise atomic clocks on fast-moving planes.

What is time?

Time is an illusion.

-Albert Einstein

Let's pause to ask an often-overlooked question: what even is time? We think of it as a constant, marching steadily forward, but it's not. Time is not a universal measure or an independent entity. Instead, time is a metric for the behaviour of energy. It measures how energy interacts and changes. When you move faster or slower through space, or when you find yourself under varying gravitational forces, the behaviour of energy, and therefore time, is altered. This is why, as you approach the speed of light, time slows down: the interplay of energy adapts to accommodate the new state of motion.

But why does motion or gravity affect time? The answer lies in the fundamental interplay between space, time, and energy. Motion and gravity influence the "fabric" of space-time, which in turn dictates how energy propagates through it. For example, in high-velocity environments, energy required to sustain motion increases exponentially as you near the speed of light. This energy distortion slows down the processes we measure as time, clocks tick slower, biological functions decelerate, and the subjective experience of time alters.

Gravity, on the other hand, warps space-time itself, creating a similar effect. Clocks positioned in stronger gravitational fields run slower compared to those in weaker fields because the intense energy density of gravity bends space-time, affecting the flow of time within it. In essence, time is the measure of how energy evolves within a given frame of space-time, not a universal constant, but an emergent property shaped by conditions.

Now consider this concept in the context of computers and video games. Time in these systems is nothing more than a programmable construct designed to optimize performance and consistency. When a game character moves quickly through a virtual world, or when a complex scene with multiple elements is rendered, the computational system often adjusts the processing priorities. For example, distant objects may be rendered at a lower resolution, or certain non-critical computations may be deferred. In a sense, the system "slows down" or reallocates resources to ensure smooth operation.

---

[16] The **t'** (read as "t prime")

Similarly, if we are living in a simulation, time dilation might be a rule programmed to regulate how processes unfold in different regions of the system. High-velocity or high-energy areas could demand greater computational resources. To optimize, the simulation could "slow down" time in these regions, balancing the load and maintaining consistency across the virtual universe. This programmed nature of time reflects how simulations balance dynamic behaviour with computational constraints.
If time is simply a measurement of how energy behaves, then we must confront the idea that time itself is not a fundamental aspect of reality. In both physics and computational systems, time emerges as a tool, a way to quantify change and regulate processes. It's not a tangible entity but a framework we use to make sense of energy's motion and interaction.

This perspective challenges the very nature of time as we know it. If time is not real but merely a measure, it aligns strikingly with the principles of a simulation. Within a simulated universe, time serves not as a fundamental constant but as a mechanism for order, variability, and resource management. And just as in video games, where "slowing down" certain actions ensures smooth performance, our universe might be doing the same, offering a programmable, optimized framework for existence.

Time, then, is not a universal truth. It is a functional illusion, a programmable feature in the grand algorithm of reality.

The distinction between past, present, and future is only a stubbornly persistent illusion.

-Albert Einstein

This statement profoundly redefines our understanding of time. If time is merely a metric for energy's behaviour, then the segmentation of events into past, present, and future is a human construct, a convenient way to organize our perception of the universe.

In reality, all moments may exist simultaneously within the framework of space-time, much like frames of a filmstrip. The "now" we experience is a slice through this continuum, determined by our motion and perspective. This aligns with the concept of block universe theory, which suggests that time is not flowing but rather laid out as a complete, unchanging structure. We perceive change and progression because of our limited perspective and how we process information.

In a simulated environment, this illusion of flowing time might serve a similar purpose: to provide a coherent narrative for the entities within it. Just as video games simulate progression with scripted sequences and timed events, our experience of time's passage might be an engineered feature, ensuring that causality and order appear consistent to the observers within the simulation. The distinction between past, present, and future, then, is not an absolute truth but a functional narrative device, a programmable artefact designed to make sense of a complex, interconnected reality.

If we imagine the universe as a box, where past, present, and future exist simultaneously, we can explore how energy and time interplay across the cosmos. From a "god-mode" perspective, viewing the universe from above, it becomes clear that the effects of energy are not constant across all regions. Gravity, motion, and space-time curvature vary significantly, causing time as we know it to tick faster in some places and slower in others. This variability means that, even from a universal vantage point, past, present, and future coexist in different parts of the box.

For example, in regions with intense gravitational forces, time slows dramatically due to space-time distortion, effectively creating pockets where the "past" lingers. Conversely, in areas with minimal gravity and high velocities, time progresses more rapidly. These differences illustrate how the fabric of space-time allows moments to exist simultaneously, depending on the subject and their position within this dynamic framework.

However, conceptualizing this simultaneity comes with inherent limits. Information itself is constrained by the speed of light, which acts as a cap on how quickly one part of the universe can interact with or know about another. Light, in essence, is a tool for moving information across space-time. This constraint introduces a delay, meaning that the effects of one region on another are not instantaneous. This is analogous to simulations and video games, where rendering distant objects or events is delayed to conserve resources and maintain performance consistency.

These limits ensure that, while the universe may appear seamless, its parts function independently to some extent. The delay in information transfer preserves order and consistency, but it also means that no single observer can fully reconcile the simultaneous existence of past, present, and future. From the god-mode perspective, time is not a flowing river but a mosaic of interdependent patches, each governed by its unique conditions and constraints.

Time Travel

Extending this perspective to the topic of time travel, we encounter a fundamental limitation: if time itself is not a free-flowing dimension but rather a quantized metric of energy's evolution, then reversing or jumping between distinct “moments” becomes exceedingly problematic. The universe, at its core, seems to operate with discrete building blocks of space-time, tiny, quantized cells that collectively give rise to the reality we perceive. Attempting to reverse the “flow” of time within a given region would require undoing the effect of energy in every one of those quantized cells simultaneously, an undertaking that scales in complexity as the region grows. Reversing a single cell's state might be marginally conceivable in extremely rare and controlled conditions, but orchestrating a reversal across an entire macroscopic region becomes, for all practical purposes, impossible.

This understanding helps clarify why some interpretations of General Relativity appear to allow time travel, yet never truly deliver the sort of backward-journey scenarios popularized in science fiction. Wormholes[17], for instance, do not genuinely transport someone to another time; rather, they function (in theory) as higher-dimensional shortcuts between remote areas of space-time. Exiting a wormhole in a distant location faster than light would otherwise allow can create the illusion of having “travelled back in time” when you look back at the place you left and see it as it was in your past (due to the light travel delay). However, you have not literally undone events or rewound the clock in those original coordinates; you are simply observing them at an earlier stage because of the finite speed of information transfer.

Hence, time travel in the traditional sense, where one could revisit a past event or leap forward to a predetermined future, clashes with the notion of a quantized, interactive, and emergent framework of time. If time is simply a measurement of energy's behaviour, then reordering those measurements is akin to asking for an impossible reconfiguration of the

---

[17] A wormhole is a hypothetical tunnel-like structure in space-time that connects two separate points in space and time, potentially allowing for faster-than-light travel. Predicted by Einstein's theory of general relativity, wormholes are often depicted in science fiction as shortcuts across the universe. Though their existence remains unproven.

entire system's underlying state. This highlights how the "time travel" allowed by certain relativistic solutions is often misunderstood and over interpreted. At best, such paths exploit geometrical properties of space-time to create shortcuts, not true temporal reversals.

Ultimately, if we accept that time is not a fundamental entity but an emergent, quantized metric of how energy unfolds, the concept of traveling "in" time becomes less about controlling a universal clock and more about navigating the information and energy constraints that shape our experience of reality. The past, present, and future may coexist in a grand, block-like tapestry of space-time, but rearranging that tapestry to one's liking remains outside the bounds of what physics, and any hypothetical simulation running it—can feasibly permit.

Length Contraction

Length contraction refers to how objects appear shorter when they move very fast relative to an observer. The equation

$$L' = L\sqrt{1 - \frac{v^2}{c^2}}$$

Explains that the moving object's length **(L′)** becomes smaller than its original length **(L)** as its speed **(v)** approaches the speed of light **(c)**. For example, if a spaceship were traveling at an incredibly high speed, it would look compressed along its direction of travel to someone watching from Earth. This phenomenon is a result of how space and time are interconnected.

Imagine that the simulation we inhabit operates with finite computational resources. Just as rendering engines in video games dynamically adjust the level of detail in visual elements to prioritize performance, the simulation might employ a similar mechanism to manage spatial representation efficiently. Length contraction, then, becomes a tool to optimize resource allocation. By dynamically altering the perceived size of objects based on their relative velocity, the simulation ensures consistent spatial relationships for observers while conserving its computational power.

Now consider the role of speed and gravity. Certain limits, such as the impossibility of massive objects reaching the speed of light, appear to be hard-coded into the simulation. Gravity inherently restricts the ability of large and heavy objects to accelerate indefinitely. This limitation not only preserves physical laws but also prevents excessive computational demands that would arise from rendering high-speed, massive entities. Even if such objects could hypothetically move at extreme velocities, their appearance and detail, from the perspective of an observer, would change. The contraction of length and the corresponding reduction in observable detail can be viewed as an optimization mechanism.

Humans, as biological observers, have inherent perceptual limits. Our eyes are incapable of tracking objects moving at extreme speeds. But it's not just human biology that imposes these limitations; even our most advanced technologies, including high-speed cameras, fail to capture objects moving beyond certain thresholds. Theoretically, an object moving at an extraordinarily high velocity could become effectively invisible, disappearing from observable reality. Why does this happen? Because at such speeds, the resources required to render the object in detail, whether to our eyes or technological instruments, would increase exponentially.

From an optimization perspective, the simulation prioritizes efficiency. By not rendering objects moving at unattainable velocities, the system conserves resources for other processes. This aligns with how video games manage elements beyond the player's immediate view, unnecessary details are omitted or simplified to focus computational power where it matters most.

Thus, length contraction may represent a deliberate optimization code within the simulation's architecture. It ensures spatial coherence while conserving resources, balancing the demands of rendering complex, high-speed phenomena against the need to maintain a stable, observable reality. The hard limits on speed and perception, whether biological or technological, are part of this optimization strategy. The question arises: if this optimization is intentional, what does it reveal about the nature of our simulated universe and the priorities of its architects?

Mass and Energy:

The law of conservation of energy, one of the most fundamental principles in physics, serves as a compelling rationale supporting the theory that we might be living in a box. This principle states that energy in a closed system remains constant over time. It cannot be created or destroyed but can only transform from one form to another.

The roots of the conservation of energy trace back to ancient Greek philosophy, particularly with thinkers like Empedocles and Aristotle. Empedocles believed that the elements (earth, water, air, fire) were indestructible and only changed forms. Aristotle built on this by discussing the concept of permanence, which implied that some qualities of nature remain unchanged. While these early ideas were not specific to energy, they introduced a foundation for the idea of enduring substances or forces within nature.

Moving to The Renaissance period which was a time of revolutionary scientific thought, laying groundwork for the conservation principle. Galileo Galilei (1564-1642) explored the nature of motion and inertia, hinting at the idea of conserved motion or energy in objects. René Descartes (1596-1650) later proposed that the quantity of motion in the universe remains constant, which was one of the earliest formulations suggesting conservation of a type of physical quantity. Descartes' work was still primitive by today's standards, but it was crucial in setting the stage for future discoveries.

In the 17th century, Christiaan Huygens expanded upon ideas of conserved quantities, specifically in relation to collisions. Huygens introduced the principle of conservation of momentum, which is related to the modern energy conservation principle. Meanwhile, Isaac Newton (1643-1727) developed his laws of motion and universal gravitation in his seminal work, which indirectly supported energy conservation concepts by establishing predictable and transferable forces in the universe. By the 18th century, the understanding of heat and motion grew significantly. The study of heat became central to the development of the conservation law, especially as it was considered a fluid-like substance called "caloric" that could neither be created nor destroyed. This caloric theory was eventually disproved, but it shaped many early ideas on energy conservation.

The 19th century was marked by rigorous experimentation and the eventual formulation of the first law of thermodynamics, stating that energy in an isolated system is conserved. James Prescott Joule (1818-1889) was a pioneer in demonstrating energy conservation through experiments showing that mechanical work could be converted into heat. His work showed a direct relationship between energy forms, culminating in Joule's Law (1845), which quantified the equivalency between heat and mechanical work. Joule's experiments laid the groundwork for a unified energy concept, marking a major milestone in understanding that energy is not substance-bound but transferable.

Other scientists, such as Hermann von Helmholtz, further developed the conservation law by applying it to biological and physical systems. In 1847, Helmholtz formally published his work on energy conservation, stating that all forms of energy mechanical, heat, electrical are interconvertible and conserved. By the mid-19th century, the conservation of energy was well established as a core principle in thermodynamics.

This principle, also known as the first law of thermodynamics, states that the *total energy within an isolated system remains constant, though it may change form.* Scientists like Rudolf Clausius and William Thomson (Lord Kelvin) contributed further to formalizing thermodynamic laws, cementing the role of energy conservation as a cornerstone of modern physics.

With the discovery of quantum mechanics and relativity, the principle of conservation was re-examined and expanded. Albert Einstein's theory of relativity introduced the concept of mass-energy equivalence, revealing that mass itself is a form of energy. This profound insight extended the conservation principle to a universal scale, demonstrating that the mass of an object can be converted into energy and vice versa. This principle governs phenomena ranging from nuclear reactions to the behaviour of subatomic particles.

The famous equation

$$E = mc^2$$

Encapsulates this idea, reinforcing the notion that energy is the fundamental currency underpinning all physical phenomena. This equation states that energy **(E)** and mass **(m)** are interchangeable, connected by the speed of light squared **(c^2).** This equation underpins the immense energy released in nuclear reactions and explains the processes that power stars, including our sun. It shows that even a small amount of mass can be converted into a vast amount of energy because **(c^2)** is an extraordinarily large number. This principle is why splitting or fusing atoms in nuclear reactions produces such massive energy.

One way to demonstrate energy conservation mathematically is through **kinetic energy** (KE) and **potential energy** (PE) in a simple closed system, like a ball falling from a height h.

- When the ball is held at height h it has potential energy:

$$PE = mgh$$

  - m = mass.
  - g = gravitational acceleration.
  - h = height.

- As the ball falls, potential energy is converted into kinetic energy:

$$KE = \frac{1}{2}mv^2$$

  - v = velocity.

- At any point during the fall, the total energy (E) of the system is:

$$E = PE + KE$$

- At the top before it falls: $E = mgh + 0$

- Just before hitting the ground: $E = 0 + \frac{1}{2}mv^2$

By using the equation of motion, we can show that $v^2 = 2gh$, so:

$$\text{KE} = \frac{1}{2}m(2gh) = mgh$$

Thus, the total energy remains $E = mgh$ at all points, proving energy conservation in this system.

Let's add some numbers to make this easier to understand.
Suppose we have a ball with a mass 2 kilograms (kg), held at a height (h) of 10 meters (m) above the ground. We'll use g = 9.8m/s^2 as gravitational acceleration.

- Mass (m) = 2kg.
- Height (h) = 10m
- Gravitational acceleration (g) = 9.8/s^2

First we calculate **Potential Energy (PE)** at the height of 10m using its formula:

$$PE = m \times x \times h$$

Substitute the values:

- First, we multiply the mass and gravitational acceleration:

$$m \times g = 2kg \times \frac{9.8m}{s^2} = 19.6N$$

Here, 19.6N is the force due to gravity acting on the ball.

- Now, multiplying this by the height (h):

$$PE = 19.6N \times 10m = 196J$$

So, the ball has **196** joules of potential energy at the top.

Now, we calculate **Kinetic Energy (KE)** just before hitting the ground. As the ball falls, this potential energy converts into kinetic energy. Just before hitting the ground, all potential energy has turned into kinetic energy.
The kinetic energy formula is:

$$KE = \frac{1}{2} \times m \times v^2$$

Where v is the velocity of the ball just before hitting the ground.
Since energy is conserved, we know that:

$$KE = 196J$$

- Let's solve for v using the KE formula:

$$v^2 = \frac{2 \times KE}{m} = \frac{2 \times 196J}{2kg}$$

- Calculating the numerator:

$$2 \times 196 = 392$$

- Divide by the mass:

$$v^2 = \frac{392}{2} = 196$$

- Take the square root to find v:

$$v = \sqrt{196} = \frac{14m}{s}$$

As you can see at the top, the potential energy is 196 J, while the kinetic energy is 0 J. Just before reaching the ground, the potential energy drops to 0 J, and the kinetic energy rises to 196 J. This constant total energy of 196 J throughout the motion demonstrates the principle of energy conservation. Here, when the ball hits the ground, its kinetic energy doesn't just disappear; instead, it transforms into other forms of energy due to the impact.

If energy cannot be created, the question arises: where did the universe's energy originate? The universe in a box theory offers a thought-provoking answer. In a simulated reality, energy might not "come from" anywhere; rather, it is a parameter set by the system's design. Much like a computer game where energy levels are predefined by code, the universe's energy could be a pre-programmed constant in a closed, virtual environment.
We live in a world governed by energy, the universal currency that facilitates every interaction, from the atomic to the cosmic scale. Humans, in this context, are not distinct entities but rather highly organized forms of energy. Death, in this framework, is the cessation of energy flow within an organism, paralleling the shutdown of a computer system when its power source is removed. This perspective aligns with the view of the universe as a "box" governed by symmetrical laws, where energy transitions but never disappears.

Every physical interaction in the universe encodes information, creating ripples that propagate through space much like signals from a Wi-Fi router. These disturbances carry intricate details about their origin, such as the position, velocity, and acceleration of objects, along with their mass and shape. Just as a router emits constant waves of data, every moment of existence, from the motion of a hand to the death of a star, generates waves of information that could theoretically travel infinitely across the universe. This encoding forms a universal record of every change, no matter how small.

For instance, the movement of an object through the air generates sound waves that reflect its motion, while subtle shifts in position create detectable changes in heat signatures. In this sense, the universe acts as a vast, continuous recorder of events, preserving a detailed history of interactions. These records might not always be accessible to us due to limitations in retrieval and the dispersal of information, but they remain embedded in the fabric of reality.

The preservation of this information finds support in quantum mechanics. Unitary evolution ensures that quantum systems retain information about their initial states within the wave function, even as it spreads or transforms. While de-coherence might make quantum information appear lost as it interacts with the environment, the larger system retains this data, emphasizing the deterministic nature of the universe at its core.

However, entropy complicates the recovery of such information, as it becomes increasingly diffused and buried within the randomness of larger systems. Still, every interaction, from particle collisions to photon emissions, leaves an indelible mark on the underlying fields of the universe. At the macro level, the universe itself offers evidence of this record-keeping. Ripples in space-time, like gravitational waves from black hole collisions, and the cosmic microwave background radiation serve as massive-scale imprints of past events. These phenomena suggest that the universe continually records and encodes its history.

Interestingly, the notion of actions being recorded aligns with concepts found in many religious and philosophical traditions. Several religions, such as Islam and Christianity, speak of a "Book of Deeds" where every human action is meticulously recorded, to be accounted for in an ultimate judgment. Similarly, Hinduism and Buddhism introduce the concept of karma, where every action, good or bad, leaves an imprint that shapes future outcomes, much like the universe's record-keeping of physical interactions. Ancient Greek philosophers, such as Heraclitus, also proposed that the universe operates on an inherent order, where actions and events ripple through time and space, leaving lasting consequences. What's astonishing is that these profound insights emerged in times when humanity had no understanding of quantum mechanics or the intricate workings of the universe. It makes one wonder: how could they have intuited such ideas about the eternal record of actions without modern scientific tools? Could it be that they were grasping at truths that science is only now beginning to uncover? or is it that they were simply clever, using logic to structure a moral system for people to behave according to their interpretation of right and wrong, while taking advantage of them by claiming superiority over them?

Consider the virtual worlds created within computers. These environments spring to life only when energy powers the machine. The digital landscapes we construct are governed by code, a finite set of instructions that determine their behaviour. Similarly, the universe exhibits traits that suggest it operates on encoded rules. From the behaviour of quarks to the laws of thermodynamics, everything seems to follow precise, predictable patterns. This symmetry and consistency hint at an underlying computational framework.

At the most fundamental level, everything in the universe, from stars to humans, is composed of quarks, governed by physical laws. These laws function like the "code" of a simulation, dictating how particles interact, combine, and form the macroscopic reality we perceive. The universe's apparent isolation, where no new energy is created, further supports the notion of a controlled system. In this light, the entire cosmos resembles a sophisticated program running within an advanced computational framework.

Free Will

To argue whether we are or are not living in a simulated reality, we must examine the concept of free will. At first glance, the choices we make may seem to reflect our autonomy. But when we look deeper, into the very fabric of reality, we find that free will is an illusion, both at the universal level and within the constraints of our lives. Everything about the nature of the universe points toward a system that is designed, optimized, and deterministic, leaving no room for true freedom of choice.

As Brian Greene aptly put it,

> *"We cannot transcend physical and mathematical laws. We are not the ultimate autonomous author of our actions. We are a collection of particles entirely governed by physical law. Every thought and action we take is nothing but those particles playing out the dictates of the mathematical equations of the physical laws."*

If the particles that make up everything in the universe lack free will, how could we, beings composed entirely of particles, possess it?

At the quantum level, particles follow strict mathematical rules. Even when they appear random, as in quantum superposition or probabilistic outcomes, their behaviour is still governed by underlying laws of physics. A photon, for example, doesn't choose whether to behave like a wave or a particle; its behaviour is dictated by the conditions of observation. An electron doesn't decide its energy level; it follows the quantized rules of its atomic environment. The principle of least action ensures that every particle takes the most efficient path, further demonstrating that particles don't act freely, they simply respond to the rules encoded in the universe.

If particles themselves lack agency, it is inconceivable that we, composed entirely of these particles, could have true free will. Our brains, thoughts, and decisions are nothing more than the interactions of countless particles obeying physical laws. Our sense of choice is an emergent property of these deterministic processes, a by-product of the system rather than a true expression of autonomy.

Not only do we lack free will on a fundamental, universal level, but we also lack it within the constraints of the universe's design. Our genetics, passed down through countless generations, shape who we are before we even have a chance to act. Our upbringing, shaped by family, society, and culture, adds another layer of restriction. Governments, societal standards, and technological systems further dictate what we can and cannot do.

These limitations are not static, they are growing. As time progresses, humans arguably have less free will than they did in the past. Early humans, for example, were bound only by the physical laws of the universe and the immediate pressures of survival. Today, however, technological, societal, and governmental systems impose increasingly complex rules. Algorithms influence what we think and believe, surveillance limits our actions, and social norms shape our choices. In this sense, time itself is a force reducing our already limited autonomy, tightening the boundaries within which we operate.

The nature of the universe itself further supports the idea that we live in a system designed for order and efficiency. Conservation laws ensure that energy and momentum are preserved, optimizing interactions. The speed of light caps information propagation, reducing complexity. Quantum mechanics resolves probabilities only upon observation, saving "resources" until necessary. Entropy drives systems toward equilibrium, and the discrete nature of particles avoids infinite detail by using quantized units. These laws are not random, they suggest a deliberate design, where constraints are necessary for the system to function. In such a designed universe, it becomes clear that free will is not a fundamental truth but an illusion. Every action, thought, or decision is governed by deterministic forces at every level, from the particles that make us up to the societal structures that guide our lives.

Whether or not we live in a simulation, the evidence overwhelmingly suggests that free will is a carefully crafted illusion. Particles, the building blocks of reality, lack agency; how, then, could beings composed of these particles possess true autonomy? Beyond this, we are constrained by genetics, societal systems, and the growing influence of technology. These limits, layered upon the deterministic structure of the universe, point to a reality where choice is not real but an emergent feature of a system designed for order. Free will, it seems, is not a fundamental property of existence but a sophisticated by-product of a universe or simulation that is optimized and controlled at every level.

Think Outside the Box

If we consider the entire universe as a box, then everything we can think, imagine, or reason is inherently limited within the boundaries of that box. Our logic, reason, and even imagination are products of this confined system. Take, for example, visual illusions or optical illusions, these demonstrate how our perception can be tricked within the rules of the universe, reinforcing the idea that our understanding is tied to the constraints of the system. Similarly, when we imagine humans with wings or the possibility of flight, this idea still aligns with the laws of this universe. It's conceivable that a form of human-like beings exists somewhere in the universe with such characteristics, but this notion does not defy the universe's underlying principles. However, we cannot conceive of ideas that exist entirely outside the box's framework, such as inventing a truly new number, envisioning a colour beyond the visible spectrum, or imagining a dimension or phenomenon that has no counterpart in our reality. These limitations underline how even our imagination, as boundless as it may feel, is ultimately restricted by the parameters of the simulated reality we inhabit.

When we speculate about what might lie outside the box, we face a fundamental challenge. If the entire universe is the box, then the "outside" would either adhere to laws similar to our universe or operate under entirely different principles. If the outside world operates under different laws, the closest we can come to conceptualizing them is to think illogically or in ways that defy the principles of reason as we understand them. Yet, even this attempt is restricted by the tools we use to think, our minds and perceptions, which are themselves products of the box. This means that even our imagination, which we often consider the ultimate expression of freedom, is confined by the boundaries of what is possible or conceivable within this system. While we may feel as though we are stretching the limits of thought, we remain tethered to the framework of the simulation.

This creates a paradox: while we may speak about the existence of an "outside," our understanding of it can never truly transcend the box. Our logic, imagination, and even our attempts to "think outside the box" are, ironically, still limited by the box's boundaries. Just as a video game character cannot comprehend the reality of its programmers, we too may be unable to fathom the nature of what lies beyond the simulation. Any theories we construct about an external reality are likely projections or extensions of concepts within our system.

For example, we might imagine advanced beings, alternate dimensions, or even higher levels of consciousness, but all these ideas are extrapolated from our existing experiences and knowledge within the box. True outside-the-box thinking, free from the constraints of the simulation, is not only difficult but may simply be impossible. Our imagination, though seemingly boundless, is in reality a product of the very boundaries it seeks to escape, forever limited by the laws and structures of the box.

The Leap to General Relativity (1915)

While Special Relativity dealt with observers moving at constant speeds, it left out an essential part of reality: gravity. Einstein's next challenge was to incorporate acceleration and gravity into his framework. This journey would take a decade of relentless work, culminating in the General Theory of Relativity in 1915.

Einstein's breakthrough came from a deceptively simple thought experiment. Imagine you're in a sealed elevator, floating in deep space. Suddenly, you feel a force pulling you to the floor. Are you experiencing gravity, or is the elevator accelerating upward? Einstein realized that there's no way to tell the difference. This insight, called the equivalence principle, suggested that gravity and acceleration are deeply connected.

But how does gravity work? Newton had described it as a force between masses, but Einstein envisioned something different. He proposed that massive objects, like the Earth or the sun, warp the fabric of space and time around them. Picture a heavy ball placed on a stretched rubber sheet: the ball creates a dip, and smaller objects rolling on the sheet will move as if pulled by an invisible force. This warping of space-time is gravity. To describe this mathematically, Einstein used the language of tensors, a branch of mathematics that deals with multi-dimensional quantities. This might be intimidating, but let's try to understand its concepts.

Einstein's Field Equation (EFE): The Foundation

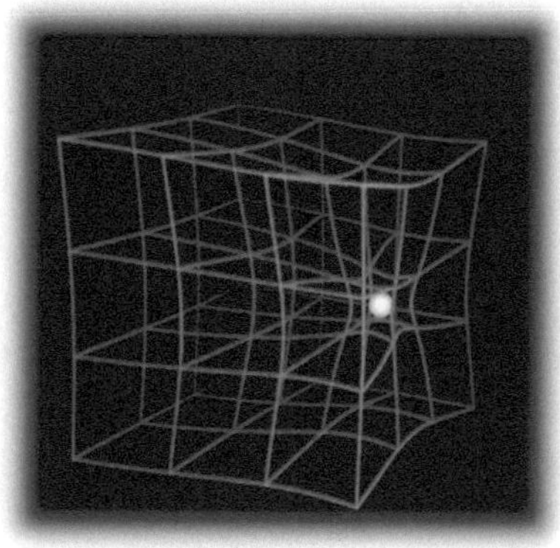

Einstein's Field Equation is the **master equation** that governs General Relativity. It describes the relationship between matter/energy and the curvature of space-time:

$$G_{\mu\nu} + \Lambda g_{\mu\nu} = \frac{8\pi G}{c^4} T_{\mu\nu}$$

This equation describes how mass and energy warp space-time. In this equation, $G_{\mu\nu}$ *"G mu nu"* (where **μ** is *"mu,"* pronounced as *"myoo"*, and **ν** is *"nu,"* pronounced as *"new"*). Represents the curvature of space-time (how "bent" space-time is), and $T_{\mu\nu}$ represents the energy and momentum of whatever matter or radiation is present. The constant **G** is the gravitational constant, and **C** is the speed of light, which appears because gravity operates within space-time. The term Λ[18]$g_{\mu\nu}$, called the cosmological constant, explains the expansion of the universe.

Essentially, this equation tells us that the presence of matter and energy determines how space-time curves, and this curvature is what we experience as gravity. For example, the sun warps space-time so that the Earth follows a curved path (its orbit) around it.

Imagine space-time as a stretched-out trampoline. When you place a heavy ball (like a bowling ball) in the centre, it creates a dip in the fabric. Smaller objects, like marbles, roll toward the heavy ball not because they're "pulled" by a force, but because the surface is curved. In the universe, massive objects like stars and planets warp space-time, creating "dips" that other objects move through, which we perceive as gravity.

Einstein's equations are remarkably general, encompassing the entire universe. However, applying them to specific scenarios, such as the behaviour of space-time near a black hole, the expansion of the universe, or the effects of time dilation near a massive object, requires further refinement and specialized solutions.

---

[18] **Λ**: Pronounced as "Lambda," representing the cosmological constant, related to dark energy or the expansion of the universe.

To address these challenges, scientists developed tailored solutions to make the Einstein Field Equations (EFE) practical for both theoretical exploration and real-world applications. While Einstein laid the groundwork, the complexity and universality of his equations demanded additional mathematical ingenuity. Pioneers like *Schwarzschild*, *Friedmann*, and *Kerr* expanded on Einstein's foundation, much like how Newtonian mechanics was later advanced by other physicists and mathematicians.

Gravitational lensing

General Relativity predicted phenomena so strange that even Einstein initially doubted them. One was the bending of light by gravity. It predicts that light bends around massive objects, a phenomenon called gravitational lensing. The angle by which light bends can be calculated using gravitational lensing equation

$$\Delta 0 = \frac{4GM}{c^2 r}$$

In this equation, $\Delta 0$ is the angle of deflection, **G** is the gravitational constant, **M** is the mass of the object causing the bending, **r** is the distance of closest approach, and **c** is the speed of light. This equation explains how massive objects like galaxy clusters distort the light from distant galaxies, creating arcs or even rings in the sky. This effect has become a critical tool in astronomy, allowing scientists to observe distant parts of the universe that would otherwise be invisible.

Picture light as a car driving on a straight highway. Now imagine the highway goes over a hill (a massive object) that causes the car to dip slightly before returning to a straight path. This bending happens because space-time is curved by the hill (the massive object). Light follows this curved path, which we observe as its "bending" around massive bodies like stars or galaxies.

This phenomenon can be likened to how rendering engines in computer graphics simulate distortions for realism, such as lens effects. The predictable way that massive objects like galaxies distort light could reflect a rule embedded in the simulation's physics engine. The fact that these distortions can reveal otherwise hidden parts of the universe, allowing us to observe objects behind massive galaxies, hints at an optimization feature, where the simulation ensures maximum visibility of its contents for observers within the system.

In 1919, British astronomer Arthur Eddington confirmed this during a solar eclipse, when starlight passing near the sun was observed to bend, just as Einstein's theory predicted. This discovery catapulted Einstein to global fame.

Gravitational wave

General Relativity also predicted black holes, regions of space-time so warped that nothing, not even light, can escape (as previously discussed), and gravitational waves. They are ripples in space-time caused by massive objects accelerating, like colliding black holes or merging neutron stars. The simplified wave equation for gravitational waves in the absence of matter is:

$$\Box \boldsymbol{h}_{\mu\nu} = 0$$

Here, $\Box$ is the *d'Alembert* operator, which describes how the wave propagates through space-time, and $\boldsymbol{h}_{\mu\nu}$ represents the small perturbations in the space-time metric caused by the gravitational waves. In simpler terms, this equation shows that gravitational waves move outward from their source at the speed of light, carrying information about the massive events that created them.

Imagine space-time as a calm pond. If you throw a heavy rock into the pond, it creates ripples that spread outward in all directions. These ripples are like gravitational waves, spreading through the "fabric" of space-time after a violent event, such as two black holes spiralling into each other. Just as the ripples in the water carry information about the size and force of the rock, gravitational waves carry information about their source, like the masses and speeds of the merging objects.

Gravitational waves aren't disturbances in a physical medium (like sound waves in air); they are distortions in space-time itself. When they pass through Earth, they stretch and squeeze space in alternating directions, incredibly subtly, by fractions of the width of an atom. Detectors like LIGO and Virgo use laser beams to measure these tiny distortions, allowing us to "hear" the universe's most cataclysmic events.

In simple terms, gravitational waves are the "echoes" of massive events in the universe, rippling through space-time like waves in a pond. They provide a new way of observing the cosmos, giving us insights into events and objects that we can't detect with light or other forms of energy. Both have since been observed: black holes through indirect evidence and gravitational waves directly detected in 2015 by the LIGO collaboration.

In a simulation, such waves could serve as data packets transmitting updates about massive events across the system. Their propagation speed, capped at **C** ensures that the simulation maintains causality and avoids inconsistencies. The detection of these waves using instruments like LIGO and Virgo, which measure minuscule distortions in space-time, aligns with the idea of a fine-tuned universe where even the smallest perturbations are governed by consistent rules. This attention to detail reflects the precision of a well-designed simulation.

Building upon (EFE), *Karl Schwarzschild*, not only formulated the concept of black holes, but also, gravitational time dilation, where time slows down near massive objects due to the curvature of space-time.

Gravitational Time Dilation

Time slows down in strong gravitational fields, a phenomenon known as gravitational time dilation, and is described by the *Schwarzschild* time dilation formula

$$t' = t\sqrt{1 - \frac{2GM}{rc^2}}$$

Here, t' represents the time experienced by someone close to a massive object, t is the time experienced far away from the object, **G** is the gravitational constant, **M** is the mass of the massive object, **r** is the distance from its centre, and **c** is the speed of light. This equation shows that the closer you are to a massive object (smaller **r**), the slower time moves for you compared to someone further away.

Think of time as a flowing river. Near a massive object (like a black hole), the river slows down, as if it's flowing through a muddy patch. Farther away, the river flows faster. This means clocks near a massive object tick more slowly compared to those far away. For example, if you lived near a black hole for a year, someone far from it might experience several years in that same period.

In a simulation, this could represent a programmed gradient, where time is not an absolute constant but adjusts dynamically based on "massive nodes" in the system. This design could optimize computational resources by slowing down processes near high-energy regions, reducing the demand on the simulation's processing capacity.

Black Holes

In a quiet room filled with chalk dust and the faint hum of intellectual curiosity, John Michell sat contemplating the cosmos. It was the late 18th century, a time when telescopes were rudimentary, and the universe was little more than a scattered array of distant lights. Yet, Michell's vision extended far beyond what his eyes could see. He asked a daring question: Could there exist stars so massive, so dense, that even light could not escape their gravitational pull?

To grasp Michell's idea, imagine trying to throw a ball into the sky. On Earth, you'd need to throw it at a speed of about 11 kilometres per second to overcome gravity, a speed known as escape velocity. Escape velocity depends on two factors: the mass of the object and its size. Now, picture a star far heavier than Earth but packed into an incredibly small space. Its density would create such intense gravity that the speed required to escape would be unimaginably high.

Michell calculated that if a star were dense enough, its escape velocity could exceed the speed of light. Since nothing travels faster than light, not even light itself could escape such a star's gravitational pull. Michell called these hypothetical objects "dark stars," invisible not because they stopped shining, but because their light was trapped by their immense gravity.

Using the formula for escape velocity, $v_e = \sqrt{\frac{2GM}{R}}$, he demonstrated that if a star's radius (R) became small enough while maintaining its mass (M), the escape velocity (v_e) could surpass the speed of light (C).

For Michell, this was a theoretical marvel. In 1783, he shared his findings, but the idea found little traction. The tools of his time were incapable of confirming such a phenomenon, and the concept faded into obscurity, a scientific footnote awaiting rediscovery.

More than a century later, general relativity was published, hinted at the possibility of extreme gravitational effects, but Einstein himself doubted their most radical implications. While Einstein remained cautious, German physicist Karl Schwarzschild took the equations further. Amid the horrors of World War I, he solved Einstein's equations for a single spherical mass, uncovering a startling result. Schwarzschild described an entity with a gravitational pull so intense that it formed an **"event horizon,"** a boundary beyond which nothing, not even light, could escape. This critical distance from the object's centre, known as the Schwarzschild radius, is given by the formula:

$$r = \sqrt{\frac{2GM}{c^2}}.$$

If an object's mass were compressed within this radius, it would become what we now call a black hole. Schwarzschild's work, though mathematically rigorous, was not fully understood at the time. Like Michell's dark stars, his findings were considered curiosities, solutions to equations rather than physical realities.

Oppenheimer-Snyder Collapse

In 1939, another monumental step was taken in our understanding of black holes when J. Robert Oppenheimer and Hartland Snyder published their ground-breaking work on the gravitational collapse of massive stars. Building on Einstein's general relativity, they provided the first modern theoretical description of how black holes could form. Oppenheimer and Snyder described a scenario in which a star, after exhausting its nuclear fuel, could no longer support itself against its own gravity. The star's core would collapse inward, compressing matter to a point of unimaginable density.

To better understand this, imagine these stars as moving through the fabric of space-time, much like a boat gliding over water. As long as they possess sufficient energy and motion, they don't sink into the fabric; they create ripples but remain buoyant. However, when a star loses its energy or the ability to sustain motion, it begins to lean heavily on the fabric of space-time. If the star's mass is great enough, this leaning effect creates a profound deformation in space-time, a hole, similar to how a stationary boat with a breach in its hull might start to sink into the river it once floated on.

When this happens to a star, the process is catastrophic. The core collapses, and layer by layer, the star compresses into an infinitely dense point called a singularity. Oppenheimer and Snyder's calculations revealed that, from the perspective of an external observer, the collapsing star would appear to freeze at the Schwarzschild radius due to the effects of time dilation. This haunting insight illustrated how the death of massive stars could create black holes, immense wells in the fabric of space-time, ushering in a new era of understanding in astrophysics.

Hawking Radiation

Fast forward to 1974, when Stephen Hawking challenged the notion of black holes as inescapable cosmic prisons. By applying quantum mechanics to the event horizon, Hawking made a startling prediction: black holes emit radiation. This phenomenon, now known as Hawking radiation, arises from quantum fluctuations near the event horizon, where pairs of particles and antiparticles, such as an electron and a positron, spontaneously form due to the energy of the vacuum. These particles are entangled, meaning their quantum properties are fundamentally linked even as they separate.

Typically, these particle pairs would annihilate each other almost immediately, returning their energy to the vacuum. However, if the pair forms very close to the event horizon, one particle can fall into the black hole while the other escapes. The escaping particle becomes part of the Hawking radiation we could theoretically observe, carrying energy away from the black hole. Meanwhile, the particle that falls into the black hole has negative energy relative to the outside universe, a concept allowed by the intense warping of space-time near the event horizon. This negative energy reduces the black hole's overall mass and energy, causing it to shrink over time.

This might lead to confusion: if one particle escapes and one falls in, wouldn't their energies cancel each other out, leaving the black hole unchanged? The answer lies in the intricate interplay of quantum mechanics and conservation laws. Before the escaping particle departs, the particle-antiparticle pair effectively imprints a copy of their quantum state onto the black hole's event horizon. This imprint acts as a kind of "ledger," preserving the information about the pair even as one particle disappears into the black hole. The black hole "pays" the energy cost for the escaping particle, ensuring that the conservation of energy is upheld, and the universe's information remains intact.

This is where the holographic principle provides a ground-breaking insight. Proposed by *Gerard 't Hooft* and further developed by *Leonard Susskind*, the principle suggests that all the information about the matter and energy that falls into a black hole is not lost but encoded on its event horizon, a two-dimensional surface that acts like a "cosmic memory." Think of the event horizon as a holographic screen, where the quantum information of everything that crosses it is stored. While the black hole's interior appears to be three-dimensional, it is essentially a projection of the information encoded on this two-dimensional boundary.

Entanglement plays a crucial role here: the escaping particle retains a quantum connection to its counterpart inside the black hole, and this connection links to the information stored on the event horizon. This mechanism suggests that black holes don't destroy information but instead transform it, encoding it on their surfaces in a way that is consistent with the laws of quantum mechanics.

As the black hole continues to emit Hawking radiation and lose mass, its event horizon shrinks. Over immense timescales, this process leads to the black hole's eventual evaporation. But what happens to the encoded information on the event horizon when the black hole vanishes? One prevailing theory is that this information is released back into the universe, subtly encoded in the emitted Hawking radiation. This concept of information recovery reconciles the black hole's evaporation with the laws of quantum mechanics, addressing the black hole information paradox.

In this way, the holographic principle bridges quantum mechanics and general relativity, transforming our understanding of black holes. Far from being voids where everything is lost, black holes are dynamic systems that act as cosmic storages, encoding and eventually releasing the energy and information they consume. They reveal a tantalizing possibility: that our entire universe might itself be a hologram, with the three-dimensional reality we perceive encoded on a distant two-dimensional boundary.

Singularity Theorem

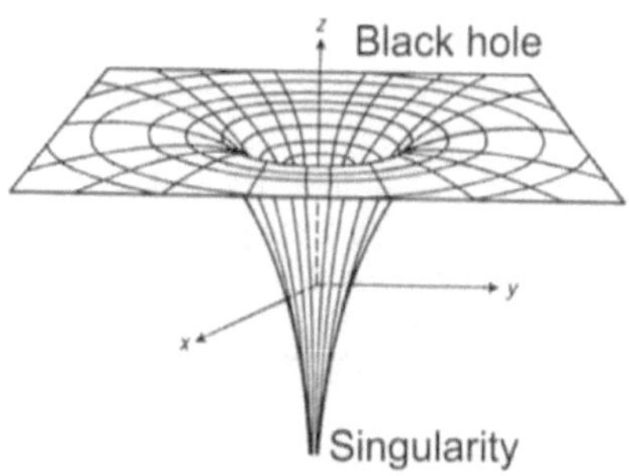

In 1965, Roger Penrose took the theoretical framework to a new level. With his seminal singularity theorem, Penrose proved that singularities, points of extreme density at the heart of black holes, are not mere artifacts of mathematical equations but inevitable outcomes of gravitational collapse under general relativity. His work relied on a deep understanding of the curvature of space-time, showing that under certain conditions, gravity overwhelms all other forces, compressing matter to an incredibly dense state.

To understand Penrose's insight, think of space-time as a stretched rubber sheet. Place a bowling ball on the sheet, and it creates a dip, representing gravity. Now, imagine increasing the ball's mass infinitely while shrinking its size. The dip becomes so extreme that it effectively tears through the sheet. This tear represents a singularity, a region where our understanding of physics, particularly Einstein's general relativity, breaks down.

However, the interpretation of singularities as infinite points of density and zero volume, a direct consequence of general relativity, may be fundamentally flawed. If true singularities with infinite density existed, they would cause such extreme distortions in the fabric of space-time that the Big Bang itself could not have occurred. Such a scenario would make the initial conditions for the universe's expansion impossible, preventing the formation of stars, planets, and the conditions necessary for life. Thus, the existence of our universe strongly suggests that the singularities described by general relativity are not infinite but instead constrained by the limits of quantum mechanics.

Classical general relativity works exceptionally well in describing space-time up to the event horizon, the boundary of a black hole. Beyond this point, however, space-time becomes so warped and extreme that general relativity no longer applies. This is because the interior of a black hole exists in a realm that is no longer part of our universe. It represents another dimension entirely, where the familiar laws of physics cease to hold, requiring new theoretical frameworks such as quantum mechanics, loop quantum gravity, or string theory to describe what happens.

Quantum theory provides a framework for understanding the limits of singularities. Unlike general relativity, which treats space-time as a smooth continuum, quantum mechanics introduces discrete, fundamental units to the structure of space-time. Loop quantum gravity, for example, suggests that space-time is composed of finite "atoms" or loops, creating a smallest possible unit of volume. Once the matter and energy within a black hole are compressed to this critical density, gravitational collapse cannot continue. Instead, the singularity reaches a finite state, governed by quantum laws.

String theory further refines this picture by suggesting that fundamental particles are not point-like but instead tiny vibrating strings. This perspective smooths out the singularities predicted by general relativity and allows for a finite core at the centre of black holes. Additionally, the uncertainty principle of quantum mechanics states that it is impossible to know both the position and momentum of particles with absolute precision, thereby preventing the collapse of matter into an infinitely small point.

Under normal circumstances, black holes maintain an equilibrium between the gravitational pull inward and the quantum effects preventing infinite collapse. This balance explains why black holes do not typically explode or collapse further. However, certain extreme scenarios could disrupt this balance, pushing the singularity past its quantum mechanical limits and causing a catastrophic event.

For a singularity to "explode" or "bounce," a specific set of conditions must occur:

1. **Exceeding the Quantum Density Threshold:** As black holes accrete mass, their internal density could approach the critical quantum limit. While this rarely happens under normal conditions, merging supermassive black holes or the collapse of an entire galaxy cluster could create a singularity dense enough to exceed this threshold. At this point, quantum effects would dominate, leading to an outward burst of energy, a quantum bounce or explosion.
2. **Infinite Energy Density in a Closed System:** In a closed space-time system, where energy density continues to increase (such as during the collapse of a region of space-time into a single point), the quantum critical limit could be surpassed. This scenario is similar to the conditions theorized before the Big Bang, where the entire universe existed in an ultra-dense, finite state that exploded outward, creating space-time itself.
3. **Primordial Black Hole Collapse at a Quantum Scale:** Tiny black holes formed in the early universe, known as primordial black holes, could evaporate through Hawking radiation. At their final stages, the remaining matter could collapse, exceed the quantum limit, and trigger a high-energy explosion resembling the Big Bang.
4. **Destabilization of a Multiverse Connection:** In certain theories, black holes might connect to other universes or dimensions. If this connection destabilizes, the accumulated matter and energy could explode, releasing energy in a manner akin to a cosmic burst.

This leads to an intriguing possibility tied to the multiverse theory: if a quantum bounce or explosion were to occur within a black hole, it could create an entirely new universe. In this scenario, the energy and matter compressed within the black hole would rapidly expand, forming a new space-time bubble within the original space-time. This "bubble within a bubble" universe would operate under its own physical laws and constants, distinct from those of the parent universe.

Such a process might explain the origin of our own universe. The Big Bang could have been the result of a quantum bounce inside a singularity in a pre-existing universe. The observed lights in our cosmos that appear to predate the Big Bang might hint at remnants of this parent universe. This theory aligns with the idea that our universe is not unique but part of an endless cycle of universe creation, each born from the collapse and bounce of singularities.

Gradually, black holes transitioned from theoretical constructs to observable phenomena. Astronomers noticed stars behaving strangely, orbiting invisible companions with immense gravitational influence. With advancements in technology, indirect evidence for black holes mounted. The Hubble Space Telescope and breakthroughs in radio astronomy revealed regions of space where gravity ruled supreme.

Then, in 2019, the Event Horizon Telescope captured the first direct image of a black hole, a glowing ring of gas surrounding a dark, empty void. This monumental achievement confirmed the existence of an object once dismissed as pure speculation. What began as John Michell's bold hypothesis and grew through Einstein's equations had finally been seen with human eyes.

The story of black holes is not merely about celestial objects; it is a testament to the extraordinary power of the human mind to transcend its boundaries. Guided by mathematics, we have ventured into the unknown, revealing truths once thought unimaginable. These equations do more than describe reality, they predict it, offering glimpses into realms beyond the reach of our current observation.

As the cosmos stretches the limits of our perception, it reveals a profound insight: while the universe may elude our eyes, it does not escape the grasp of our minds. Mathematics stands as the bridge between the visible and the invisible, connecting what we know to the mysteries we have yet to uncover.

This is a tribute to the transformative power of theoretical physics. By pursuing mathematical elegance, pioneers in the field unveiled truths about the universe that were astonishingly ahead of their time. Their work demonstrated that mathematics is not only a tool for describing the physical world but also a means of predicting phenomena that had never before been observed.

The Expanding Universe

Dark Energy

Dark energy is one of the greatest mysteries in modern cosmology, representing the force driving the accelerated expansion of the universe. Its discovery and implications have fundamentally transformed our understanding of the cosmos, linking it to key theoretical developments, such as the Friedmann equation, and observational breakthroughs like Hubble's Law.

Friedmann equation

The theoretical groundwork for understanding the universe's expansion began with Karl Schwarzschild's pioneering work in the early 20th century, which used Einstein's General Relativity to describe space-time around massive objects. Building on this foundation, Alexander Friedmann extended General Relativity in the 1920s to describe the dynamics of the entire universe. Friedmann's equations revealed that the cosmos is not static, as previously assumed, but dynamic, expanding or contracting depending on the interplay of matter, energy, and the mysterious force we now call dark energy.

Friedmann's work culminated in the formulation of the Friedmann equation, a cornerstone of modern cosmology:

$$(\frac{\dot{a}}{a})^2 = \frac{8\pi G}{3} p - \frac{k}{a^2} + \frac{\Lambda}{3}$$

Describes how the universe expands over time. In this equation, **a** represents the scale factor, which measures the size of the universe at any given moment, while $\dot{a}$ is its rate of change, showing how fast the universe is expanding. The term **ρ** corresponds to the density of matter and energy in the universe, and **k** determines the curvature of the universe, whether it's flat, open, or closed. Lastly, **Λ**, the cosmological constant, accounts for dark energy, the mysterious force responsible for the universe's accelerated expansion.

To understand this better, imagine the universe as a balloon with small dots drawn on it, representing galaxies. As you blow up the balloon, the dots move farther apart, not because they are moving across the balloon's surface but because the surface itself is stretching. Similarly, galaxies aren't flying through space; instead, space itself is expanding, carrying them apart.

This equation shows that the rate of expansion depends on the interplay between the gravitational pull of matter and energy (which tries to slow the expansion) and the repulsive force of dark energy (which accelerates it).

However, the Friedmann equation also allows for the possibility of contraction under certain conditions. If the density of matter and energy is sufficiently high and the curvature of the universe is closed **(K >0)**, the gravitational pull could overcome the expansive forces, causing the scale factor ($\dot{a} < 0$) to decrease over time.

Similarly, if the cosmological constant (Λ) were negative, it would exert an attractive force that could lead to the universe contracting instead of expanding. This scenario, known as the "Big Crunch," envisions a future where the universe reverses its expansion, eventually collapsing in on itself. While current observations suggest that dark energy drives an accelerating expansion, the *Friedmann* equation reminds us that the universe's fate depends on the delicate balance of these forces, leaving the door open to the possibility of contraction.

Hubble's law

Friedmann's prediction was confirmed observationally decades later when Edwin Hubble discovered in 1929 that galaxies are moving away from us, as indicated by their redshift.

Hubble's work, calculates the speed at which an astronomical object, such as a galaxy, is receding from us due to the expansion of the universe.

Hubble's Law

$$v = H_0 \, . \, d$$

Demonstrated that the universe is expanding and introduced H_0 Hubble's constant, a key measure of this expansion. Here **v** represents the recession speed of the object, H_0, is the Hubble constant (the rate of expansion of the universe, typically measured in kilometres per second per mega parsec), and **d** is the distance to the object, usually measured in mega parsecs. The equation demonstrates a linear relationship, meaning the farther an object is, the faster it appears to move away. For instance, if the Hubble constant is 70 km/s/Mpc, an object 10 mega parsecs away will have a recession speed of 700 km/s.

This relationship helps astronomers quantify the universe's expansion by linking observed redshifts (indicators of how fast an object is moving away) to the distances of celestial objects. However, when scientists measure H_0 today, they get conflicting results depending on the method used. Observations of the early universe, such as from the Cosmic Microwave Background, yield a lower value (~67 km/s/Mpc), while measurements of nearby galaxies give a higher value (~73 km/s/Mpc). This discrepancy, known as the **Hubble tension.**

The Hubble Constant: A Misleading Constant

To understand why the interpretation of Hubble's Law might be flawed and why the Hubble constant is not truly constant, we must first delve into the nature of dark energy and its workings.

Dark energy can be thought of as a property of space-time itself, an intrinsic energy that permeates the cosmos, the underlying background of the universe. Think of space-time and dark energy as two sides of the coin. Imagine space-time as a vast ocean, with planets, stars, and galaxies as objects floating on its surface. In regions dominated by matter and energy, gravity creates a curvature in space-time, akin to a heavy ball deforming a stretched sheet. This curvature generates gravitational forces that act to counteract the expansive effects of dark energy.

In less dense regions, such as cosmic voids where gravitational forces are weak or negligible, dark energy dominates, driving the expansion of space-time at an accelerating rate. Thus, the balance between gravity and dark energy varies across the universe, influencing how space-time behaves locally and on a cosmic scale.

Dark energy behaves in some ways like biological cells. Under specific conditions, cells multiply rapidly, but their growth slows or ceases when constrained by environmental forces. Similarly, dark energy's influence expands faster in regions with minimal gravitational resistance, while in areas of strong gravitational forces, its expansion slows or halts entirely. This analogy helps illustrate how dark energy's "multiplication" or its accelerating expansion is context-dependent, shaped by its environment.

The universe expands isotopically, meaning it stretches uniformly in all directions. This might lead one to assume that the rate of expansion is identical everywhere, but the reality is more nuanced.

At the edges of the universe, the expansion speed is uniform because these regions are devoid of significant matter, energy, and gravitational forces. With no opposing forces to counteract dark energy, the fabric of space-time expands at its fastest possible rate in these areas. Importantly, this expansion is not driven by a force acting from one side or another; rather, it occurs everywhere simultaneously, much like the way biological cells multiply uniformly under favourable conditions.

Within the universe, however, the situation is different. The expansion rate varies because matter, energy, and gravitational forces interact with the space-time fabric in complex ways. Regions dense with galaxies and other matter experience slower expansion due to gravity's counteracting influence. In contrast, less dense areas, such as cosmic voids, expand more rapidly as dark energy becomes dominant. Thus, while the universe expands uniformly on a large scale, the local rates of expansion are influenced by the distribution of matter and energy within it.

This phenomenon raises an important question: Why doesn't this expansion affect local systems like galaxies, planetary systems, or even Earth itself? The answer lies in the overwhelming strength of local gravitational forces. In areas where matter is densely packed, such as within galaxies or solar systems, gravity binds objects together tightly enough to resist the expansive force of dark energy. As a result, the space-time fabric does not stretch significantly on these smaller scales, preserving the stability of local systems.

In essence, while dark energy drives the universe's expansion on the largest scales, its effects are negligible in regions dominated by gravitational forces, allowing local systems to remain intact. This duality highlights the complex interplay between gravity and dark energy, shaping the universe in ways that vary dramatically depending on scale and context.

The accelerating expansion of the universe is supported by several key observations:

- **Distant Supernovae:** Measurements of *Type Ia supernovae* reveal that their redshifts, caused by the stretching of light wavelengths as space expands, are greater than expected. This indicates that the expansion of the universe is speeding up over time.
- **Cosmic Microwave Background (CMB):** The CMB provides a snapshot of the early universe, offering clues about its large-scale structure and the influence of dark energy on its evolution.
- **Large-Scale Structures:** Observations of galaxy clusters and voids further confirm that dark energy constitutes approximately 68% of the universe's total energy density.

A common misconception is that the universe's expansion is limited by the speed of light. In reality, the expansion of space-time itself is not constrained by this limit. While no object with mass can travel faster than light relative to its surroundings, the fabric of space-time, driven by dark energy, can stretch faster than light over vast distances.

Hubble's Law effectively describes the motion of objects in gravitationally dense regions, but it does not account for the full variability of expansion rates across the universe. Since the Hubble constant is derived from measurements in gravitationally dense areas, such as galaxies and galaxy clusters, it primarily reflects the **slowest expansion rates**, where gravity counteracts dark energy's influence. This means it does not represent the rapid expansion occurring in less dense regions like cosmic voids, where dark energy dominates and space-time stretches more freely.

Interestingly, the density of dark energy remains constant, for now. Imagine the universe as a box, with dark energy as its fabric. The current expansion hasn't reached the box's boundaries. Once those boundaries are reached, the density of dark energy will begin to increase. Currently, galaxies are moving apart because dark energy's density isn't strong enough to overcome local gravitational forces. However, as its density increases, even galaxies within clusters will start to separate from each other locally.

The evidence overwhelmingly supports the interpretation of dark energy as a dynamic field, not as a constant in its behaviour. While dark energy exists uniformly across the universe, its influence, how it drives the expansion of space-time, is not static but varies depending on local and cosmic conditions. This perspective is both logically consistent and better aligned with observational evidence.

A constant dark energy, as assumed in the traditional cosmological constant **(Λ)** model, would imply that its effect is uniform across all regions and times, exerting an unchanging repulsive force on space-time. However, this is contradicted by key observations.

Logically, the concept of dark energy as a dynamic field aligns with the nature of other cosmic forces and phenomena. Gravity, for instance, is not uniform in its effects; it varies with mass and distance. Similarly, the expansion of the universe itself is not isotropic on smaller scales, influenced by the distribution of matter and energy. It would be inconsistent to assume that dark energy, which interacts with and is shaped by the same cosmic fabric, would behave uniformly everywhere. A dynamic field, such as Quintessence, evolves in response to its environment, becoming stronger in areas with minimal gravitational interference and weaker in regions where gravity dominates.

To better understand the distinction, imagine dark energy as the "wind" driving the expansion of a sailboat across a vast ocean (the universe). If the wind were constant, the boat would always move at the same speed regardless of the ocean's conditions. However, in reality, the wind (dark energy) interacts with varying ocean currents (gravitational forces), sometimes accelerating the boat in calm waters (cosmic voids) and slowing it in rougher seas (gravitationally dense regions). This behaviour reflects the dynamic nature of dark energy's influence, even though the "wind" itself permeates the entire ocean.

Observational evidence also supports this dynamic behaviour. *Type Ia* supernovae, the cosmic microwave background (CMB), and the distribution of galaxy clusters all reveal that the expansion of the universe accelerates more in some regions than in others, depending on the interplay between dark energy and local gravitational forces. These findings cannot be explained by a constant dark energy model, which would predict a uniform expansion rate unaffected by local variations.

Therefore, we can modify the Friedmann equation by replacing the cosmological constant $\boldsymbol{\Lambda}$ with a term that represents the energy density of a scalar field $\boldsymbol{\phi}$, which is dynamic.

The energy density and pressure of the scalar field are given by:

$$P\phi = \frac{1}{2}\dot{\phi}^2 + V(\phi)$$

$$p\phi = \frac{1}{2}\dot{\phi}^2 - V(\phi)$$

Here, $\frac{\dot{\phi}^2}{2}$ is the kinetic energy of the scalar field and $V(\phi)$ is the potential energy of the scalar field.

Hence, the equation becomes:

$$\left(\frac{\dot{a}}{a}\right)^2 = \frac{8\pi G}{3}\left(P_{m}a^{-3} + P_{r}a^{-4} + P_{\phi}\right) - \frac{k}{a^2}$$

Where P_{ϕ} is dynamically governed by the field equations.

The dynamics of the scalar field ϕ are determined by the Klein-Gordon equation in an expanding universe:

$$\ddot{\phi} + 3H\dot{\phi} + \frac{\partial V}{\partial \phi} = 0$$

Here, $H = \frac{\dot{a}}{a}$, is is the Hubble parameter and $\frac{\partial V}{\partial \phi}$ is the derivative of the potential $\mathbf{V(\phi)}$, which captures how the scalar field evolves dynamically in time.

Dark energy as a dynamic scalar field evolves in response to the expansion of the universe and its own potential $V(\phi)$. Unlike a constant Λ, this approach allows for time-varying behaviour of dark energy, such as transitioning between dominating and non-dominating phases, depending on the shape of $V(\phi)$. Hence, the final modified formulation would be:

$$\left(\frac{\dot{a}}{a}\right)^2 = \frac{8\pi G}{3}\left(P_{m}a^{-3} + P_{r}a^{-4} + \frac{1}{2}\dot{\phi}^2 + V(\phi)\right) - \frac{k}{a^2}$$

Cyrus law

Modern cosmology typically describes the universe's expansion using the Friedmann-Lemaître-Robertson-Walker (FLRW) metric, which assumes a uniform distribution of matter and energy at large scales. This approach leads to a single global expansion rate described by the Hubble parameter. However, the real universe exhibits significant inhomogeneities galaxy clusters, filaments, and cosmic voids, and potentially subtle variations in dark energy. These structures suggest that the local expansion rate may differ from the global average. With this in mind, we introduce the "**Cyrus Law**": a phenomenological framework designed to capture how local matter density and dark energy variations influence the local expansion rate, thereby bridging the gap between global and local cosmic dynamics.

Cyrus Law posits that the universe's local expansion rate is not strictly uniform but varies regionally based on local matter density and dark energy properties. In over-dense regions, where gravitational fields are stronger, the expansion rate is suppressed, whereas under-dense regions (voids) can maintain or even exceed the baseline expansion rate. When averaged over large volumes, these local variations reveal a more nuanced global expansion parameter, reflecting the universe's complex structure and evolution.

Although not yet an established cosmological law, the Cyrus Law and its corresponding equation provide a structured framework for understanding how local cosmic conditions influence expansion rates. With further theoretical refinement and rigorous validation through simulations and observational data, it could become a valuable tool in elucidating the intricate interplay between matter distribution, dark energy, and global cosmic acceleration.

Motivation

Inhomogeneity of Matter:

Observations show that matter is unevenly distributed. Galaxy clusters form in over-dense regions where gravity slows the expansion locally. Conversely, large voids contain far less matter, allowing local expansion to proceed more freely.

Dark Energy Variations:

While standard cosmology treats dark energy as uniform, it is conceivable that subtle variations or effective local properties could exist. Such variations may locally enhance or diminish the influence of dark energy on expansion.

Bridging Scales:

Traditional cosmological parameters like the Hubble constant are global averages. Cyrus Law seeks to provide a formalism to move from local conditions to a meaningful global rate that inherently reflects the universe's complexity

Mathematical Framework

$H_\Lambda(t)$**:** The baseline (or homogeneous) expansion rate driven by dark energy in a smoothly distributed universe (often associated with a cosmological constant **(Λ).**

$p_m(x,t)$ **:** Local matter density at position x and cosmic time **t**.

$\bar{p}_m(t)$ **:** Cosmic mean matter density at the same time.

$\delta(x,t) = \frac{p_m(x,t)-\bar{p}_m(t)}{\bar{p}_m(t)}$: The local matter density contrast. Over-dense regions have $\delta > 0$, under-dense regions have $\delta < 0$.

$\Phi(x,t)$: A dimensionless spatial scaling function that representing how local variations in dark energy, curvature, or other physics modify the expansion. If $\Phi \approx 1$, no exotic local enhancement is present. If $\Phi > 1$, local expansion may be boosted; if $\Phi < 1$, it may be diminished.

Define positive constants **k** and **B** that set the sensitivity and non-linearity of the local expansion rate to matter density variations. These parameters would be chosen based on theoretical arguments and tuned by comparison with cosmological simulations or observations.

Cyrus Equation:

$$H(x,t) = \frac{H_\Lambda(t)\ \Phi(x,t)}{1 + k\,[\delta(x,t)]^B}$$

When $\delta > 0$ (over-dense region), the denominator $1 + k\,[\delta(x,t)]^B$ becomes large, reducing $H(x,t)$ below $H_\Lambda(t)$ Local gravitational fields slow expansion here.

When $\delta \leq 0$ (under-dense region), the denominator approaches unity. If $\Phi(x,t)$ is close to 1, the local expansion rate remains near the baseline. If $\Phi(x,t) > 1$ it can exceed the baseline, reflecting freer expansion in voids.

Thus, the Cyrus Equation encapsulates the idea that local conditions (captured by δ and Φ) determine the local expansion rate relative to a uniform baseline.

Global Expansion Rate:

To obtain a meaningful global expansion parameter, we average $H(x,t)$ over a large representative cosmic volume **V**:

$$H_{global}(t) = \frac{1}{V}\int_V H(x,t)dV = H_\Lambda(t)\langle\frac{\Phi(x,t)}{1 + k\,[\,\delta(x,t)]^B}\rangle$$

If the universe is statistically homogeneous and isotropic on large scales, this average yields a global expansion rate naturally influenced by the spectrum of over-densities, voids, and any spatial variations in dark energy. Unlike a single Hubble constant $H_{global}(t)$ emerges from an integral that inherently acknowledges cosmic structure.

The Cyrus Law is a phenomenological model. It does not replace Einstein's field equations or standard cosmological perturbation theory; rather, it supplements them by providing a convenient parameterization of local expansion effects. It is inspired by known principles:

- **N-body Simulations**: Large-scale simulations already show that local environments differ in their expansion dynamics. The Cyrus Equation can be tested against such simulations, fine-tuning **k**, **B**, and forms of $\mathbf{\Phi(x,t)}$ to match numerical results.
- **Inhomogeneous Solutions**: Exact but complex inhomogeneous cosmological solutions (e.g., Lemaître-Tolman-Bondi models) demonstrate how different regions can evolve at different rates. The Cyrus Equation is a simplified, "effective" approach to capture similar behaviour.

Implications:

Testing and Calibration:

- By comparing the Cyrus Law predictions with outcomes from high-resolution simulations, one can determine whether a particular choice of **k**, **B**, and $\mathbf{\Phi(x,t)}$ realistically reproduces observed large-scale structures.

Observational Signature:

- Detailed measurements of the Hubble parameter in various directions, gravitational lensing patterns, galaxy cluster observations, and void studies could, in principle, detect subtle scale- or direction-dependent variations in measured expansion rates. Such findings would provide observational support for or constraints on the Cyrus framework.

Refining $\mathbf{\Phi(x,t)}$:

- Further theoretical exploration might yield physically motivated forms of $\mathbf{\Phi(x,t)}$. For example, it might relate to variations in the local equation of state of dark energy or track environmental effects on cosmic acceleration.

Proof

The "Cyrus Law" as presented is a phenomenological model. A rigorous "proof" from first principles (i.e., directly from Einstein's field equations for general inhomogeneous universes) is non-trivial and beyond the scope of a simple presentation. Instead, we will demonstrate that under reasonable cosmological assumptions and approximations, the Cyrus Equation reproduces known behaviour of local expansion rates in inhomogeneous universes and reduces correctly to standard linear perturbation results.

Consider a universe described on large scales by the **FLRW** metric:

$$ds^2 = -c^2dt^2 + a^2(t)[\frac{dr^2}{1 - k_r r^2} + r^2(d0^2 + sin^2 0\, d\phi^2]$$

Where **a(t)** is the scale factor and k_r is the spatial curvature parameter (not to be confused with the parameter **k** in the Cyrus Equation). The global Hubble parameter is defined as $H(t) = \dot{a}(t)/a(t)$. When the universe is dominated by a cosmological constant $\mathbf{\Lambda}$ and a smoothly distributed matter component, we have a baseline expansion rate $H_\Lambda(t)$**.** In the exact homogeneous case:

$$H_\Lambda(t) = \sqrt{\frac{8\pi G}{3}\,(\bar{p}_m(t) + P\Lambda)}$$

Where $\bar{p}_m(t)$ is the mean matter density, and $P\Lambda$ is the dark energy density associated with **Λ.**

In reality, matter is not uniformly distributed. We define a local density field $p_m(x,t)$ and a density contrast:

$$\delta(\mathrm{x},\mathrm{t}) = \frac{p_m(x,t) - \bar{p}_m(t)}{\bar{p}_m(t)}$$

For small perturbations **(|δ| ≪ 1)**, linear perturbation theory provides a well-established approach to study structure growth. In Newtonian gauge, scalar perturbations to the metric lead to local gravitational potentials **Ψ (x, t)** that affect local expansion rates.

In linear theory and under Newtonian approximation, the local expansion can be approximated by:

$$H\,(x,t) \approx H_\Lambda(t)\,(1 - a\delta\,(\mathrm{x},\mathrm{t}))$$

For some proportionality **α** related to how gravitational potentials slow down local expansion in over-dense regions. Although **α** depends on details of the perturbation growth and gravitational potential, the key takeaway is that to first order, local expansion rates decrease linearly with increasing density contrast.

This standard linear result suggests two conditions that a more general phenomenological law should satisfy in the weak perturbation limit:

- **Correct Baseline Limit**: When $\delta(\mathrm{x},\mathrm{t}) \to 0$, we must have $H\,(x,t) \to H_\Lambda(t)$.
- **Monotonic Decrease for Over-densities**: For **δ > 0** and small, $H\,(x,t)$ should decrease approximately linearly with **δ.**

Considering the Cyrus equation, Let us consider the linear regime, where **|δ| ≪ 1 and Φ (x, t)** is close to unity **(Φ (x, t) ≈ 1.**

Expand the denominator for small **δ:** $1 + k\,[\delta\,(\mathrm{x},\mathrm{t}]^B \approx 1 + \mathrm{k}\delta\,(\mathrm{x},\mathrm{t})$ if we choose **B** =1 for simplicity in the linear limit. Thus:

$$H\,(x,t) \approx \frac{H_{\Lambda\,(t)}}{1 + k\delta(\mathrm{x},\mathrm{t})}$$

For **|k δ (x, t) | ≪ 1**, we can use the approximation $\frac{1}{1+x} \approx 1 - \mathrm{x}$:

$$H\,(x,t) \approx H_{\Lambda\,(t)}\,(1 - \mathrm{k}\delta\,(\mathrm{x},\mathrm{t}))$$

This matches the general form predicted by linear perturbation theory, where local over-densities reduce the local expansion rate linearly. By identifying **α = k**, we see that the Cyrus Equation reduces correctly to the known linear form for small perturbations. This shows internal consistency and that it can reproduce standard results in the appropriate limit.

Now consider a scenario where **Φ (x, t)** slightly deviates from unity, say **Φ (x, t) = 1 + ϵ (x, t)** with **|ϵ| ≪ 1**. In the linear regime:

$$H\,(x,t)\;\approx\;\frac{H_{\Lambda\,(t)}\,(\mathbf{1}+\boldsymbol{\epsilon}\,(\mathbf{x},\mathbf{t})}{\mathbf{1}+\,\mathbf{k}\boldsymbol{\delta}\,(\mathbf{x},\mathbf{t})}$$

Expanding:

$$\mathrm{H(x,t)}\approx H_{\Lambda\,(t)}[(\mathbf{1}+\boldsymbol{\epsilon}\,(\mathbf{x},\mathbf{t}))(\mathbf{1}-\mathbf{k}\boldsymbol{\delta}(\mathbf{x},\mathbf{t}))].$$

To first order:

$$\mathrm{H(x,t)}\approx H_{\Lambda\,(t)}[\mathbf{1}+\boldsymbol{\epsilon}(\mathbf{x},\mathbf{t})-\mathbf{k}\boldsymbol{\delta}(\mathbf{x},\mathbf{t})].$$

This means that if **Φ (x, t) > 1**, it can partially offset the decrease due to positive δ, potentially allowing under-dense regions δ < 0 to have **H (x, t) ≥** $H_{\Lambda\,(t)}$ This is exactly the qualitative behaviour we expect: under-dense regions expand more freely, and small positive **ϵ** epsilonϵ can reflect enhanced local expansion due to effective "extra push" from local dark energy variations or reduced gravitational inhibition.

For larger **δ** (beyond the linear regime), the chosen exponent **B** controls the non-linearity. Higher values of **B > 1** increase the penalty for large over-densities, causing a more pronounced suppression of **H(x,t)** in regions of significant matter concentration. While a complete non-linear "proof" would require solving the Einstein field equations for a given inhomogeneous distribution, the Cyrus Equation’s form is motivated by this logical extension of the linear limit: it encapsulates both the linear regime and provides a tunable non-linear response.

Finally, to connect back to a global expansion rate. If perturbations are small on average or distributed symmetrically, and **⟨δ (x, t)⟩ ≈ 0**, the global rate simplifies to something close to $H_{\Lambda\,(t)}$. Deviations in higher moments of **δ** and specific distributions of **Φ** produce a statistically corrected global expansion rate that naturally incorporates structure formation. This approach is consistent with how cosmological perturbation theory integrates local fluctuations into a global picture.

While not a rigorous proof derived strictly from Einstein’s field equations, the Cyrus Equation demonstrates its validity by recovering the correct linear perturbation behaviour in the limit of small **δ**, aligning with the expectation that over-dense regions suppress local expansion rates while under-dense regions can maintain or exceed the baseline rate when **Φ > 1**. Furthermore, it provides a framework for averaging to a global expansion rate that inherently accounts for the effects of inhomogeneities. These attributes establish that the Cyrus Law is not arbitrary but is phenomenologically anchored to well-understood linear approximations, serving as a tractable extension that captures non-linear effects and spatial variations in dark energy.

The Universe's ultimate fate: Reset

The Big Crunch

The Big Crunch is a theoretical scenario in which the expansion of the universe ultimately reverses due to the force of gravity. We already know that the Big Bang happened, an event that marked the beginning of our universe as we understand it. This leads us to excogitate that a Big Crunch may have already occurred in the past. If this is true, it is entirely possible that the laws of the universe in that previous iteration were different.

To understand this, we must first examine the concept of critical density, the threshold that determines the universe's ultimate fate. Critical density is defined as the precise density of matter and energy required for the universe to neither expand forever nor collapse. This concept was first derived by physicists in the mid-20th century, notably by William H. Press and Paul Schechter in the 1970s, building upon earlier cosmological models inspired by the works of Einstein and Edwin Hubble. Their analysis was motivated by the need to reconcile observed galactic distributions with the large-scale dynamics of the universe.

Its formula is:

$$P_c = \frac{3H^2}{8_\pi G}$$

Where:

- P_c: Critical density
- H: Hubble constant (rate of universe's expansion)
- G: Gravitational constant

If the actual density of the universe is greater than this critical density, the attractive force of gravity could slow the expansion and eventually reverse it. This reversal could culminate in a catastrophic collapse where galaxies, stars, and matter are drawn together into a singular point (singularity), mirroring the conditions of the Big Bang.

At present, the universe is expanding, but could this change? It would be logical and highly possible that there is a limit to the expansion of the universe. Since we know the universe cannot be static, once the expansion limit is reached, the universe may stop growing. At this point, gravity could slowly regain dominance, potentially leading to the Big Crunch. One key factor in this process could be gravitational over-density, which refers to regions of space where matter is densely concentrated, amplifying gravitational pull and potentially tipping the scales toward collapse.

The Big Freeze

If the Big Bang was a reset caused by a Big Crunch, it's possible that gravity, the dominant force in the earlier universe, played a key role in that collapse. However, not all theories predict an ultimate collapse. Instead, the universe might face a Big Freeze scenario.

In this model, the universe continues expanding indefinitely. As it does, galaxies drift farther apart, and the density of matter and energy decreases. Over immense timescales, stars burn out, and black holes evaporate through *Hawking radiation*. Without energy sources, the universe becomes a cold, dark, and lifeless expanse.

The Big Freeze is closely tied to the concept of entropy, which measures disorder. The second law of thermodynamics states that entropy in a closed system tends to increase. As the universe expands and energy spreads thinly across an ever-larger space, it approaches a state of maximum entropy, where no usable energy remains.

To illustrate, consider a cup of hot coffee in a cold room: over time, the coffee cools, and the energy within it disperses, leaving everything at the same temperature, a state of equilibrium. Similarly, the universe in a Big Freeze reaches thermal equilibrium, where all processes cease.

However, this does not preclude the possibility of a reset. As the expansion continues, there could be a tear in the fabric of space-time, potentially triggering an explosion reminiscent of the Big Bang. Energy, while dissipated, may reorganize or reconfigure itself in ways we do not yet understand. Each reset could involve the emergence of new laws of physics, as seen in the Cyclic Universe Models, suggesting that the conditions leading to the Big Freeze might differ from those that led to previous cycles. Since a Big Crunch occurred in the earlier universe, it's highly possible that the new set of laws formed after the reset now favour something else, illustrating how the universe's fate can vary with each cycle.

The Big Rip

Another compelling theory regarding the universe's ultimate fate is the Big Rip. The Big Rip is a captivating hypothesis that proposes an ultimate end to the universe, where all matter, from galaxies to atoms, is torn apart due to the relentless expansion of the cosmos. This theory stands as one of the strongest possible fates of the universe, alongside the Big Crunch and the Big Freeze.

The concept of the Big Rip was first formally introduced in a 2003 paper by physicists Robert R. Caldwell, Marc Kamionkowski, and Nevin N. Weinberg. This theory emerged from earlier studies on the accelerated expansion of the universe, a phenomenon first discovered in the late 1990s by two independent teams of astronomers led by Saul Perlmutter, Brian Schmidt, and Adam Riess.

The discovery was ground-breaking, as it revealed that the universe's expansion, driven by dark energy, was not slowing down as previously assumed but was instead accelerating. Building on this discovery, Caldwell and his colleagues explored a hypothetical form of dark energy called **phantom** energy, which has profound and dramatic implications for the universe. To understand phantom energy, we must first introduce the equation of state in cosmology. This equation relates the pressure **(P)** of a fluid to its energy density **(p)** and is defined as:

$$w = \frac{P}{p}$$

Here, P is the pressure exerted by the energy, and p is the energy density. For **phantom** energy, $w < -1$, indicating an exceptionally strong negative pressure. This negative pressure is significant because it drives the accelerated expansion of the universe at an increasingly rapid rate over time. Unlike most forms of energy, where density decreases as the universe expands, the energy density of phantom energy grows. This violates the usual conservation laws of energy and causes unbounded growth in the universe's scale factor (a).

The energy density p of phantom energy evolves according to the equation:

$$p \propto a^{-3(1+w)}$$ [19]

Since w<−1, the factor −3(1+w) becomes positive. This means that as the scale factor (a) grows, (p) increases rather than decreases. This unbounded growth in energy density drives the expansion that defines the Big Rip.

The dynamics of the universe under the influence of phantom energy are governed by the *Friedmann equations*, which describe how the universe expands based on its energy content. For a spatially flat universe (k=0), the first Friedmann equation is:

$$H^2 = \frac{8_{\pi}G}{3} p$$

In this equation:

- **H** = $\frac{\dot{a}}{a}$ is the Hubble parameter, which represents the rate of expansion of the universe.
- **a** is the scale factor, describing the size of the universe relative to an earlier time.
- **G** is the gravitational constant.

In the presence of phantom energy, p (the energy density) increases over time instead of diminishing. This leads to an ever-growing H, meaning the rate of expansion accelerates exponentially. As the energy density of phantom energy grows, its effects become increasingly dramatic, eventually resulting in the Big Rip.

In this scenario, the relentless growth of energy density not only tears apart galaxies and stars but also disintegrates planets and atomic particles, destroying the very fabric of the universe. Interestingly, Quantum Field Theory supports the concept of phantom energy with its exotic properties, such as a negative equation of state parameter (w <−1). Meanwhile, the second law of thermodynamics aligns with the increasing energy density of phantom energy over time, driving the system toward higher, and ultimately destructive, energy states.

Caldwell and his colleagues outlined a timeline for the Big Rip:

I. **60 Million Years Before the Big Rip**: At this stage, galaxies will begin to drift apart as the gravitational bonds holding them together weaken due to the accelerated expansion of the universe.

II. **10 Million Years Before the Big Rip:** Galaxy clusters will lose cohesion, and individual galaxies will be separated entirely.

III. **3 Months Before the Big Rip:** The destruction of solar systems occurs. Stars, planets, and other celestial bodies will no longer be gravitationally bound, as the expanding universe overwhelms these forces.

---

19 The symbol ∝ is pronounced "proportional to" and indicates that one quantity varies in direct relation to another. For example, in this equation, p changes in proportion to $a^{-3}(1+w)$.

IV. **30 Minutes Before the Big Rip**: Stars, planets, and other structures are torn apart into smaller fragments. At this stage, the repulsive force of phantom energy grows overwhelmingly strong.

V. **Final Moments Before the Big Rip**: The molecular bonds within matter break, and even subatomic particles such as protons and neutrons are disintegrated. The universe becomes a sea of elementary particles devoid of structure or order.

So, what if the universe were not flat? What if it were closed (with k=+1) or open (with k=−1)? Would any of these scenarios change the outcome or timing of the Big Rip?

To answer this, we need to consider how the curvature of the universe interacts with phantom energy. In the *Friedmann* equation, the curvature term $\frac{-k}{a^2}$ becomes significant early in the universe's evolution, when the scale factor (a) is small. For a closed universe (k=+1), this term acts as a decelerating force, slowing down the expansion. Conversely, for an open universe (k=−1), the term contributes positively to the expansion rate. However, as the universe evolves and the scale factor grows, the impact of the curvature term diminishes because it is proportional to $\frac{1}{a^2}$, which becomes negligible compared to the rapidly increasing energy density (p) of phantom energy.

Phantom energy dominates the dynamics because its energy density grows exponentially as $p \propto a^{-3(1+w)}$ where w<−1. This growth completely overwhelms the effects of curvature. For both closed and open universes, as well as flat universes (k=0), the same outcome occurs: the scale factor grows without bound, leading to the consequences of the Big Rip.

Therefore, we could argue that the universe might have a closed curvature (k= + 1) based on the growth rate of dark energy over time. In the early universe, the growth rate of dark energy was relatively low, as matter and radiation dominated the dynamics. This slower growth rate allowed the curvature term $\frac{-k}{a^2}$ to have a more noticeable impact on the universe's expansion.

In a closed universe, this term acts as a decelerating force, creating conditions that could explain why the expansion of the universe was less accelerated during its earlier stages. As time progressed and the scale factor (a) increased, the influence of the curvature term diminished, and dark energy, specifically phantom energy, began to dominate. The observed increase in the growth rate of dark energy could then be interpreted as the natural progression of a universe transitioning from being curvature-dominated to dark-energy-dominated, which aligns with the behaviour expected in a closed universe.

This perspective offers a potential explanation for why the expansion of the universe appears to have shifted from slower early growth to the current era of rapid acceleration, consistent with the influence of a closed curvature.

The timing of the Big Rip remains primarily dependent on the equation of state parameter (w) and the current density of dark energy p_{dark}, rather than the curvature of the universe. In their original paper, Caldwell and his team estimated that a Big Rip could occur approximately 22 billion years from now, assuming **w < −1** and a current dark energy density consistent with observational data from cosmic microwave background measurements and supernova surveys.

Imagine the universe as a car driving on an endless road, with its speed representing the rate of expansion. The "engine" powering this expansion is dark energy, and its behaviour is described by the equation of state parameter, (w), which is the ratio of pressure (P) to energy density (p).

Currently, the influence of dark energy is growing, but its density is not increasing. This means the universe's expansion is accelerating, but it hasn't yet hit the theoretical "borders" or limits of the universe. If the density of dark energy were to increase, it would imply the universe has definite borders, and when those borders are reached, expansion would not stop, but the density and pressure of dark energy would stabilize, potentially leading to a new cosmic phase.

Now, here's the key relationship: when (w) decreases (becomes more negative, like going from (−1 to−1.1), it means the pressure from dark energy grows stronger compared to its energy density. This stronger negative pressure accelerates the car even faster, causing the universe's expansion to speed up uncontrollably.

This is the "phantom energy" scenario, where the rapid growth of pressure could lead to a "**Big Rip**", tearing apart galaxies, stars, planets, and even atoms. Conversely, if (w) increases (moves closer to 0, like from −1 to −0.9), the pressure weakens relative to energy density. In this case, the expansion might slow, stabilizing the universe temporarily or even reversing into a contraction.

However, we know a static universe, a universe that neither expands nor contracts, is not physically possible. A shift of (w) closer to 0, with weakened pressure, would eventually allow gravity to dominate, leading to the collapse of the universe into a "Big Crunch." Thus, (w) acts as a "dial" that determines how strongly dark energy accelerates the universe's expansion or how it might transition into a future collapse.

Today, with $w \approx -1$, we are in a state of steady acceleration. The current observations suggest that the universe's expansion is not bound by limits, its density isn't increasing yet, but the influence of dark energy continues to grow. The exact future depends on whether (w) shifts more negative, leading to runaway expansion, or closer to 0, signalling an eventual collapse. Either way, the universe's destiny lies in the interplay between dark energy's pressure, density, and w.

Now that we know the effect of w, let's recalculate the timing of the Big Rip using $w < -1.1$ with the following formulary:

$$t_{rip-}\, t_0 = \frac{2}{3 \mid 1 + w \mid H_0 \sqrt{\Omega_{DE}}}$$

Where:

H_0 is the current Hubble constant.

Ω_{DE} is the fraction of dark energy density in the universe.

w is the dark energy equation of state parameter.

Assuming:

$$H_0 \approx \frac{\frac{70\text{km}}{s}}{\text{Mpc}} = 2.27 \times 10^{-18} s^{-1}$$

$$\Omega_{DE} \approx 0.7$$

$$w = -1.1$$

$$t_{rip} - t_0 = \frac{2}{3 \times 0.1 \times 2.27 \times 10^{-18}\sqrt{0.7}}$$

$$t_{rip} - t_0 = \frac{2}{1.91 \times 10^{-19}} \approx 1.05 \times 10^{19} s$$

$$t_{rip} - t_0 \approx 1.05 \times 10^{19} s \div (60 \times 60 \times 24 \times 365) \approx 3.3 \text{ Billion years.}$$

When comparing the Big Crunch, Big Freeze, and Big Rip scenarios, each represents a distinct possibility for the universe's ultimate fate. The Big Crunch envisions a collapse driven by gravity, reversing the current expansion and potentially leading to a reset. The Big Freeze, in contrast, describes an unending expansion where entropy maximizes, resulting in a lifeless, static universe. The Big Rip stands apart by suggesting a dramatic and violent tearing apart of all matter due to the ever-increasing force of dark energy. Either way, the universe as a whole is on a path to eventually destroy itself and begin a new cycle, whether through the Big Crunch, the Big Freeze or the Big Rip.

Among these, the Big Rip is currently considered the most probable based on observational evidence of accelerated expansion. The critical factor lies in the equation of state parameter, which appears to suggest conditions favourable to a Big Rip. Unlike the Big Crunch, which relies on the density of the universe surpassing critical density, or the Big Freeze, which assumes dark energy remains constant, the Big Rip's dependence on intensifying dark energy makes it a more imminent threat if the current trajectory persists.

Schrödinger Equation (1926) Quantum Mechanics

By the mid-1920s, the world of physics was shrouded in mystery and contradiction. Classical mechanics, with its clear-cut laws and predictable outcomes, had begun to unravel under the strain of discoveries at the atomic scale. Particles behaved like waves, and waves acted like particles. Energy was no longer a continuous stream but came in discrete packets. At the heart of these puzzles was a profound question: What rules governed the smallest building blocks of nature?

The journey to resolve this question was far from straightforward. The groundwork was laid by Max Planck and Albert Einstein, whose daring ideas about quantized energy and the dual nature of light challenged conventional wisdom. Experimental breakthroughs revealed the atom, once thought indivisible, to be a complex structure of a dense nucleus surrounded by electrons. Yet, these electrons defied explanation, stubbornly resisting the predictions of classical physics.

In 1924, *Louis de Broglie*, a French physicist, proposed a bold idea: if light could behave as both a particle and a wave, perhaps matter could do the same. He introduced the concept of matter waves, suggesting that particles like electrons had an associated wavelength given by the formula:

$$\lambda = \frac{\mathbf{h}}{p}$$

Where **(λ , *lambda*)** is the wavelength, **(h)** is Planck's constant, and **(p)** is the momentum of the particle. De Broglie's hypothesis was later confirmed experimentally, setting the stage for Schrödinger's work.

It was into this intellectual ferment that *Erwin Schrödinger* stepped, profoundly influenced by Louis de Broglie's revolutionary suggestion. An Austrian physicist with a flair for the abstract and the mathematical, *Schrödinger* imagined electrons not as tiny particles zipping around the nucleus but as wave-like entities existing in a cloud of probabilities. His ambitious quest was to craft a mathematical framework that could describe these strange wave-particles not as contradictions but as a unified phenomenon. What emerged from his efforts was nothing short of a scientific masterpiece: the Schrödinger Equation. Formulated in 1926, this equation, particularly in its time-dependent form, is:

$$i\hbar \frac{\partial \psi}{\partial t} = \hat{H}\psi$$

This fundamental equation describes how the quantum state of a physical system evolves over time. In this expression, **(i)** is the imaginary unit, emphasizing the wave-like nature of particles and the intrinsic connection between quantum mechanics and complex numbers. The reduced Planck constant, **($\hbar$)**, ("h-bar"), is a fundamental constant that sets the scale of quantum effects.

The term $\frac{\partial \psi}{\partial t}$ ("partial psi partial t"), represents the rate of change of the wave function **(ψ)** (psi), over time. The wave function is a mathematical representation of the system's quantum state, containing information about properties like position, momentum, and energy. Importantly, the square of its magnitude, ($|\psi|^2$), provides the probability density of finding a particle in a specific location. The operator **($\hat{H}$)**, known as the Hamiltonian, encapsulates the total energy of the system, including kinetic and potential energy. Acting on the wave function, it dictates how the energy of the system influences its evolution.

To visualize this, imagine a boat floating on water. The boat represents a particle, while the water's ripples reflect the particle's wave-like nature. Rather than being confined to a single, fixed point, the boat becomes part of the ripples, which represent probabilities of where the boat might be. Schrödinger's equation acts as a map, predicting how these ripples evolve over time and guiding us in understanding both the particle's position and its movement.

This equation encapsulates the probabilistic and wave-like behaviours of particles, forming the foundation for phenomena like superposition, tunnelling, and wave-particle duality. Max Born's interpretation added a key insight: the square of the wave function's magnitude, ($|\psi|^2$), represents the probability density of finding a particle in a specific location. For instance, a high, ($|\psi|^2$), at a point indicates a greater likelihood of the particle being found there.

This probabilistic nature marked a dramatic shift from classical physics, where the future position and momentum of an object could be precisely determined. In quantum mechanics, only probabilities can be calculated, reflecting the inherently uncertain and fascinating world of the quantum.

The Schrödinger Equation can be likened to a musical score. Just as a score describes how notes evolve to create a melody, the equation describes how the wave function evolves to dictate the behaviour of particles. The wave function's oscillations are like the vibrations of a violin string, resonating in specific patterns that correspond to quantized energy levels.

Solving the Schrödinger Equation

To make the equation practical, Schrödinger solved it for specific systems, starting with the hydrogen atom. When applied to hydrogen, the solutions yielded discrete energy levels, matching Bohr's earlier model but with far greater theoretical rigor. These energy levels arose naturally from the mathematics, revealing the quantized nature of atomic systems.

For example, in the case of a particle in a one-dimensional box (a simple model system), the wave function solutions are sinusoidal, and the energy levels are given by:

$$E_n = \frac{n^2 \pi^2 \hbar^2}{2mL^2}$$

Here, E_n is the energy associated with the quantum number **(n)**, which is a positive integer (1,2,3,...) that determines the discrete energy levels of the particle. The term **($\hbar$)** ("h-bar") is the reduced Planck's constant, **(m)** is the mass of the particle, and **(L)** is the length of the box.

This result demonstrates the quantized nature of energy in confined systems, meaning the particle can only possess certain discrete energy values depending on its quantum state. The energy increases with the square of **(n)**, indicating that higher quantum states correspond to higher energy levels. This quantization arises purely from the mathematical constraints of solving the Schrödinger equation, offering a profound insight into the behaviour of quantum systems. The particle's confinement in the finite space of the box leads to standing wave solutions, with the wavelength of these waves directly linked to the particle's energy.

Imagine a particle, like an electron, trapped inside a small box. At first, you might picture it bouncing around like a tiny ball, but in the quantum world, particles don't act like that. Instead, they behave like waves. This wave doesn't mean the particle is physically waving like water or sound. Rather, the wave represents the likelihood of where the particle might be if we looked for it. The higher the wave at a certain point, the more likely the particle is to be found there.

So, what makes this wave form? It's the particle's energy and its confinement within the box. The particle is constantly in motion due to its energy, but because it's confined by the walls of the box, it can only "move" in specific patterns, forming what we call standing waves. Think of these as ripples that fit neatly inside the box. The waves are a result of the particle interacting with the space it's confined to.

Now, let's talk about energy. In the quantum world, energy doesn't work like a continuous flow. The particle's energy can only exist in certain fixed amounts, or "steps," similar to how a computer only understands 0s and 1s. But unlike the binary choices of a computer, the particle has more possibilities, represented by whole numbers called quantum numbers (1, 2, 3, and so on). Each quantum number corresponds to a specific amount of energy that the particle can have. For example, at the lowest quantum number, the particle has the least energy, and as the quantum number increases, so does the energy, like climbing higher rungs on a ladder.

When we say the particle's energy "increases" or that it "takes on specific values," we mean that its wave changes shape. A higher quantum number means the wave has more peaks and troughs, fitting more "ripples" into the box. These ripples represent the different states of energy the particle can occupy. It's not about the particle vibrating physically, but rather how it responds to and interacts with energy in its confined space. At lower quantum numbers, the particle's wave is simpler, with fewer ripples. At higher numbers, the wave becomes more complex, like higher-pitched notes on a guitar string.

Finally, let's address the size of the box. If the box is small, the waves are squeezed tightly, meaning the particle has higher energy levels because it has less room to spread out. In a larger box, the waves are more spread out, leading to lower energy levels. The size of the box controls how tightly the particle's wave can "fit" inside it.

Before these discoveries, science thought of particles like electrons as tiny, solid objects that could move freely in any way. Schrödinger's work changed that view entirely. It showed that particles are not just tiny objects but also behave like waves, and their energy isn't continuous, it's quantized, meaning they can only exist at specific levels. This shift in understanding revealed that the universe operates in a much more structured and mathematical way than we ever imagined. It means that the behaviour of atoms, which make up everything around us, follows precise rules.

Wave-particle duality

The Double-Slit Experiment: The Light and Shadow of Superposition

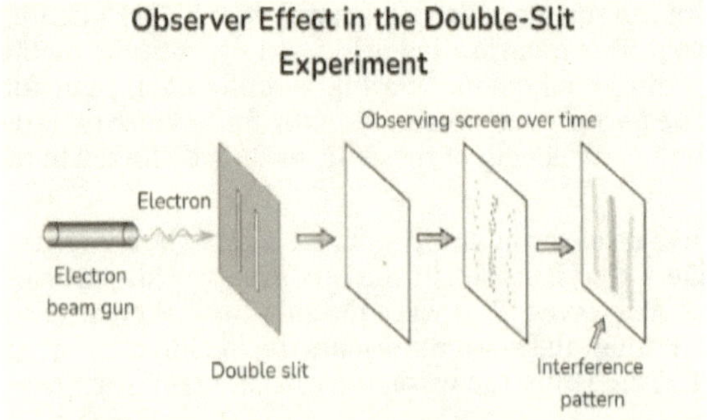

The double-slit experiment revolutionized our understanding of quantum mechanics and the nature of reality. Initially performed with light in the 1800s, the experiment was later repeated in the 20th century with electrons, leading to astonishing revelations about the wave-particle duality of matter.

Imagine spraying water through two small slits cut into a board. On the other side, you'd expect to see two streams of water matching the slits. But what if, instead, you observed a series of ripples, like those on a pond? This analogy illustrates the unexpected outcome of the double-slit experiment.

When light passes through two slits, it creates an interference pattern of bright and dark fringes on a screen behind, a behaviour expected of waves. Scientists then conducted the experiment with electrons, treating them as tiny particles. They expected electrons to pass through one slit or the other, forming two distinct bands on the screen.

What actually appeared was not two distinct bands but an interference pattern, similar to that of waves. This suggested the electrons were behaving like waves, passing through both slits simultaneously and interfering with themselves. This phenomenon defied classical logic, where particles should take only one path.

The experiment's implications deepened when electrons were fired one at a time. Even in this scenario, the interference pattern emerged after many electrons had been sent through. This indicated that each electron somehow interfered with itself, as though it travelled through both slits at once, a hallmark of quantum superposition.

To further test this strange behavioürs, scientists placed detectors at the slits to observe which path the electron took. The moment they measured which slit the electron passed through, the interference pattern vanished, and the electrons behaved like particles, creating two distinct bands. Observing the system forced the electron to "choose" a path, collapsing its superposed state.

This confirmed that superposition is a real phenomenon. Particles exist in multiple states simultaneously, represented by a wave function, until they are **observed**. The act of measurement collapses the wave function into a single state.

Superposition

Schrödinger's Cat: A Paradox in a Box

To further illustrate the implications of superposition, in 1935, Schrödinger proposed a thought experiment to illustrate the strange implications of quantum mechanics, particularly the concept of superposition.

The Scenario: A cat is placed in a sealed box with:

- A radioactive atom.
- A Geiger counter to detect the atom's decay.
- A vial of poison that breaks if the Geiger counter is triggered.

If the atom decays (a random quantum event), the Geiger counter releases the poison, killing the cat. If the atom doesn't decay, the cat remains alive.

According to quantum mechanics, the radioactive atom exists in a superposition of decayed and not decayed until observed. This means the cat, tied to the atom's state, is both alive and dead simultaneously, a macroscopic superposition. Only when the box is opened does the cat's state collapse into one reality: alive or dead. Schrödinger's cat highlights the unsettling notion that reality is not definite but probabilistic, determined by observation.

To grasp these concepts, consider flipping a coin and covering it before looking. Intuitively, the coin is either heads or tails; you just don't know which. In the quantum world, however, the coin exists as both heads and tails simultaneously until observed.

The Double-Slit Experiment: Shows how particles like electrons behave as waves, existing in all possible paths (superposition) until observed. Schrödinger's Cat: Extends this principle to macroscopic objects, illustrating the unsettling idea that observation determines reality.

The principle of superposition is mathematically described by Schrödinger's equation. A particle's state is represented by its wave function **(ψ),** which is a combination of all possible states:

$$\psi = c_1\psi A + c_2\psi B$$

Here, c_1 and c_2 are coefficients representing the probability amplitudes of each state. When a measurement is made, the wave function collapses into one state, with probabilities given by $|c_1|^2$ and $|c_2|^2$.

Both the double-slit experiment and Schrödinger's cat challenge our classical understanding of reality. They reveal a universe that is not fixed and deterministic but fluid and probabilistic, shaped by observation. At its core, quantum mechanics suggests that the universe may not "exist" in a definite form until observed, a revolutionary idea that forces us to rethink the relationship between the observer and reality.

What Determines the Collapsed State?

The collapse of the wave function, a transition from a state of superposition to a single definitive outcome, is central to quantum mechanics. Different interpretations offer varying perspectives on what determines this collapse:

Copenhagen Interpretation

Proposed in the 1920s by Niels Bohr and Werner Heisenberg, the Copenhagen Interpretation posits that a quantum system exists in superposition until it is observed. Observation forces the system to "choose" a state probabilistically, based on the squared magnitudes ($|c_1|^2$, $|c_2|^2$) of its wave function components. The observer and measurement apparatus play an integral role in defining reality, though why this happens remains unexplained. This interpretation aligns with experimental results, such as the double-slit experiment, but leaves the mechanism of collapse a mystery.

Many-Worlds Interpretation

Introduced by Hugh Everett in 1957, the Many-Worlds Interpretation rejects the idea of collapse. Instead, all possible outcomes of a quantum event occur, each in a separate, branching universe. When a measurement is made, the observer becomes part of this branching process, experiencing only one outcome in their universe. While elegant in its avoidance of randomness, this interpretation is difficult to test experimentally due to the inaccessibility of parallel worlds.

Pilot Wave Theory (Bohmian Mechanics)

Developed in the 1950s by David Bohm, this theory introduces a deterministic framework where particles are guided by a "pilot wave." The wave evolves according to quantum mechanics, while the particle's position determines the observed outcome. Hidden variables within the wave guide the system's behaviours. Though Pilot Wave Theory aligns with experimental results, its reliance on nonlocality and hidden variables has sparked debate.

Quantum Bayesianism (QBism)

Emerging in the 21st century, QBism reframes quantum mechanics as a tool for updating subjective probabilities. The wave function represents the observer's knowledge about a system, and measurement updates this knowledge rather than collapsing reality. This interpretation emphasizes the role of the observer but does not require physical collapse, focusing instead on the informational aspect of quantum mechanics.

A Universe Without Observers

If observers did not exist, quantum mechanics suggests the universe might remain in a state of indefinite potentiality. Superposition would prevail, with no collapse to actualized states. This raises profound questions: Is the universe inherently tied to observation, or does it independently manifest outcomes?

The idea that reality is "observer-centric" hints at a universe seemingly designed for consciousness. Without observers, it could appear static, unresolved, a potentiality rather than actuality. This aligns intriguingly with simulation theory, suggesting our perceived reality might be a computed reflection of a deeper, unobserved primary reality. If a simulation optimizes resources, only rendering details when observed (like video game graphics), it mirrors the observer-dependent behaviours seen in quantum mechanics.

In modern video games, environments and events are rendered only when the player approaches, optimizing computational resources. Similarly, the universe may operate on a "need-to-render" basis, collapsing wave functions only when observation necessitates it. This analogy supports the idea of an efficient, resource-conscious system, whether a simulated reality or a universe with built-in observational mechanisms.

Quantization, the discretization of energy levels, space, and time, is a hallmark of quantum mechanics and also underpins digital computation. In computers, data is quantized into bits, facilitating efficient processing and storage. Similarly, the universe's quantization might reflect an underlying digital structure, suggesting it operates as a computational entity. This necessity for quantization could ensure stability, predictability, and efficiency, aligning with the idea of a simulated framework.

The universe's design appears to allow for all possibilities, maintaining flexibility while adhering to probabilistic laws. This ensures adaptability while operating within a defined framework. The apparent unpredictability of quantum events might be an illusion of scale, at a larger level, all possibilities may be predetermined or confined to a vast, yet finite, structure. This duality of openness and constraint suggests an intelligent design, whether natural or simulated, ensuring the system can adapt to unforeseen conditions without chaos.

The convergence of quantum mechanics, observer-centric principles, and simulation theory challenges the notion of a primary, objective reality. Instead, what we perceive may be a projection or reflection of a deeper truth. Like shadows cast on a wall, our experiences could be representations, not the ultimate form of existence. Whether a product of natural laws or deliberate simulation, the universe's design hints at profound interconnectedness, where observation, quantization, and possibility converge to create the tapestry of reality.

Quantum Entanglement

Quantum entanglement is one of the most intriguing and counterintuitive phenomena in physics, fundamentally reshaping our understanding of reality. The concept was first introduced in 1935 by Albert Einstein, *Boris Podolsky*, and Nathan Rosen in a paper that became known as the EPR Paradox. They aimed to highlight a perceived flaw in quantum mechanics, which predicted that two particles could become "entangled" in such a way that measuring one would instantaneously determine the state of the other, regardless of the distance between them. Einstein, sceptical of this interpretation, famously referred to it as "spooky action at a distance." Around the same time, Erwin Schrödinger coined the term "entanglement" to describe this peculiar quantum behaviour.

To understand entanglement, imagine a pair of dice. Before either is rolled, the dice exist in a mysterious state called "superposition," where all six possible outcomes coexist. When one die is rolled and lands on, say, a 4, this act of measurement forces it to "choose" an outcome. If the dice are entangled, the roll of one die instantly determines the outcome of the other, no matter how far apart they are. For instance, if one die lands on 4, the other might always land on 3. This analogy mirrors the behaviour of entangled quantum particles, such as electrons or photons, whose properties, like spin or polarization, are intrinsically linked.

When two particles become entangled, measuring one determines the state of the other in a complementary way. For example, if one particle's spin is measured as "up," the other's spin will always be "down." Quantum mechanics explains this using the wave function, a mathematical description of all possible states a quantum system can occupy. Entangled particles share a single wave function, meaning their individual states are inseparable until one is measured, causing the wave function to collapse and the state of both particles to be determined simultaneously.

Quantum entanglement challenges the principle of locality because measurements on one particle seem to instantly affect the state of the other, regardless of the distance between them. This instantaneous "non-local" correlation appears to violate the principle that information cannot travel faster than the speed of light.

However, this interpretation can be misunderstood or misrepresented. Consider the particles in their superposition state: the argument is that they are in superposition because we, as observers, don't know their outcomes. When we measure one particle, we know the outcome of the other, which makes it seem as though the second particle instantaneously falls into its state. However, until the second particle is measured, it remains in a pre-measurement state, not yet collapsed into a definite outcome. In this sense, the entangled system as a whole remains in a pre-measurement state, and the act of measurement does not directly violate locality, it simply reveals the correlations already encoded in the shared wave function.

Einstein and his colleagues proposed an alternative explanation to avoid this perceived violation of locality. They argued that entangled particles carried hidden variables, pre-determined information about their states, encoded at the moment of their creation. It's as if the particles exchanged a secret letter containing all possible outcomes, and when one particle is measured, the information in the letter reveals the result.

Importantly, the secret letter analogy doesn't imply that all outcomes are fixed in a deterministic sense. Instead, the letter could also encode probabilistic tendencies, ranges of likelihood for certain states, while still allowing for some level of randomness in the particle's final behaviour. In this view, the hidden variables themselves could include probabilistic elements, much like the wave function does. The letter might specify probabilities for various outcomes and leave certain aspects to be resolved by the particle at random, adhering to the encoded probabilities.

This interpretation suggests that the hidden variable theory and the wave function are not entirely at odds. Both describe a reality where outcomes are not strictly predetermined but emerge through a combination of encoded probabilities and randomness. The key difference lies in the mechanism: the hidden variables represent localized information already embedded in the particles, while the wave function relies on non-local correlations that appear to transcend classical causality. Thus, the "secret letter" analogy bridges some conceptual gaps, showing that hidden variables might incorporate probabilistic elements similar to those in quantum mechanics, albeit with a classical, localized underpinning.

Bell's Theorem

In 1964, physicist John S. Bell developed a framework to test whether hidden variables or quantum mechanics accurately described reality. His work resulted in the derivation of the CHSH inequality, a mathematical expression that sets constraints on the correlations between entangled particles under local hidden variable theories. The CHSH inequality involves two entangled particles, measured by two observers, Alice and Bob. Each observer can choose between two measurement settings:

Alice: $\boldsymbol{A_1 \; or \; A_2}$, Bob: $\boldsymbol{B_1 \; or \; B_2}$

The outcomes of these measurements are represented as either +1 or −1. The correlation between Alice's and Bob's measurements is defined by the correlation function:

$$E\,(A, B) = \langle A \cdot B \rangle$$

Where **E (A , B)** is the average product of the measurement outcomes for a specific pair of settings.

The CHSH inequality can then be expressed as:

$$S = |\ \mathrm{E}\,(A_1, B_1) + E\,(A_1, B_2) + E\,(A_2, B_1) - E\,(A_2, B_2)\,| \leq 2$$

Here:

- **S** is the Bell parameter.
- **E (A, B)** represents the correlation for a specific pair of measurement settings.
- The value 2 is the maximum allowed under local realism (the assumption that information cannot travel faster than the speed of light and that physical properties exist before measurement).

Quantum mechanics predicts that entangled particles can exhibit correlations that violate the CHSH inequality. Specifically:

$$S_{quantum} \leq 2\sqrt{2} \approx 2.828$$

This maximum value is achieved when the measurements are made at specific angles, such as 45° apart for polarization measurements of photons. This violation of the inequality demonstrates the non-local nature of quantum mechanics and rules out local hidden variable theories.

The theoretical predictions of Bell's theorem were put to the test in a series of experiments conducted by physicists *John F. Clauser*, Alain Aspect, and *Anton Zeilinger*, who were awarded the Nobel Prize in Physics in 2022 for their contributions.

- **John F. Clauser**: In 1972, Clauser and Stuart Freedman conducted the first experimental test of Bell's inequality using entangled photons. They measured the polarization correlations of photon pairs and found that the results violated Bell's inequality, providing strong evidence against local hidden variable theories and supporting the predictions of quantum mechanics.

- **Alain Aspect**: In the early 1980s, Alain Aspect and his team enhanced John Clauser's earlier experiments by introducing rapid and random changes to the measurement settings while the photons were in flight. This innovation addressed what is known as the "locality loophole", a potential issue in earlier experiments where the settings of the measurement devices could theoretically influence the source of the photons if they were not sufficiently separated in space and time. By ensuring that the choice of measurement settings was made independently and during the photons' travel, Aspect's work eliminated this possibility. His experiments demonstrated even more compelling violations of Bell inequalities, providing stronger evidence for the non-local nature of quantum entanglement.
- **Anton Zeilinger**: Zeilinger's work extended the scope of entanglement experiments by demonstrating entanglement over long distances and developing quantum communication protocols. His experiments demonstrated phenomena such as entanglement swapping, where two previously un-entangled particles become entangled through their interaction with a third pair of entangled particles, and quantum teleportation, where the state of one particle is transferred to another without any physical transmission. These breakthroughs laid the foundation for practical quantum technologies, including quantum cryptography and quantum networks.

Zeilinger's work also introduced the concept of an entangled network, where particles become interconnected such that the state of one particle instantaneously influences the state of another, regardless of distance. In an entangled network, the emphasis is on the deep interconnection and dependency among the nodes (or elements) of the network, where the state or action of one node has immediate and potentially profound impacts on others. This concept is not only theoretical but has practical applications in fields like quantum computing, distributed systems, and even social and biological networks.

To understand this, imagine a spider web. Each point where threads intersect represents a node in the network. When one part of the web vibrates, the effect is felt throughout the structure. In a traditional network, the vibration might travel along predictable paths, based on the web's connections. However, in an entangled network, it's as if touching one thread instantaneously affects all others, regardless of their position. This reflects the unique non-local property of quantum entanglement, where the relationship between nodes defies the classical constraints of distance or direct interaction.

For a real-world example, consider block-chain technology, where transactions are recorded across a distributed ledger. Although this isn't quantum-based, it mimics the interdependence of an entangled network. A single change or addition to the ledger updates every copy of the block-chain simultaneously. In a quantum internet, which aims to leverage true quantum entanglement, communication could be instantaneous and immune to traditional eavesdropping, fundamentally altering the way data is transmitted.

While these experimental results are widely interpreted as ruling out any local hidden variable explanation, a more careful look at both the theoretical structure of Bell's inequalities and the experimental methodologies suggests that this conclusion may not be as airtight as it appears.

Bell's theorem rests on several core assumptions. Chief among them is the independence of measurement settings (often referred to as "measurement independence" or "free will"), which stipulates that the choice of measurement angles or settings cannot be influenced by the hidden variables themselves. If, however, the universe, at a fundamental level, allows for correlations between the hidden variables and the experimental settings, Bell's inequality can be violated without invoking non-locality. This possibility, sometimes referred to as "superdeterminism," might appear philosophically unpalatable, yet it highlights that Bell's theorem is only as robust as its assumptions.

Another subtlety lies in the common assumption that hidden variables must predetermine outcomes in a fully deterministic sense. But a hidden variable model could incorporate intrinsic probabilities, encoded in local "instructions" that remain unknown until measurement, rather than strict, predetermined values. In this framework, the hidden variables specify the probability distributions of possible measurement outcomes, operating much like the wave function but in a local manner. Bell's formalism was designed primarily to target classical, deterministic hidden variable theories (or their straightforward probabilistic extensions), leaving open the question of whether more sophisticated or higher-dimensional formulations might evade its constraints.

On the experimental side, significant progress has been made in closing various loopholes (locality, detection, and freedom-of-choice, among others). Nonetheless, strict closure of all loopholes simultaneously remains challenging. Even the most advanced tests rely on assumptions about how detectors function, how random the selection of measurement settings truly is, and how space-time intervals separate the measurement events. Tiny inefficiencies or correlations in random number generators could, in principle, allow hidden variable models to mimic a Bell violation.

Moreover, experiments are often finely tuned to align with quantum theory's predictions. While that's excellent for testing quantum mechanics under controlled conditions, it also means that any hidden variable approach seeking to pass these same tests would need equally precise fine-tuning. One could argue that these highly optimized scenarios may not represent a definitive "universal" test but rather a specialized context in which quantum theory triumphs. A more general or varied set of experiments, with different types of particle sources, detectors, and measurement protocols, could reveal subtler clues about local realistic models, especially if they incorporate probabilistic hidden variables or additional dimensional frameworks.

String theory, one of the leading candidates for a unified physical theory, introduces the idea that our four-dimensional space-time could be embedded in a higher-dimensional manifold. Phenomena that appear "non-local" in our 3+1 dimensional world may, in fact, be local correlations looping through extra dimensions. This perspective opens the door to hidden variable interpretations that reside in higher-dimensional spaces, where information travels or becomes entangled through pathways invisible in our lower-dimensional slice of reality.

In this view, Bell-type violations might not require instantaneous "spooky action at a distance" but rather result from a deeper geometric structure that connects events via dimensions beyond our immediate perception. Such frameworks could reconcile quantum entanglement with a broader concept of locality, effectively embedding local hidden variables in these extra dimensions, thereby rendering them non-local only from the standpoint of the observer trapped in 3+1 dimensions.

Perhaps the most provocative direction is to consider that the wave function and hidden variables are not mutually exclusive descriptions but partial glimpses of a unified underlying reality. The wave function captures the statistical and non-local correlations that quantum mechanics has successfully described for nearly a century. Meanwhile, hidden variables, in a higher-dimensional, probabilistically enriched sense, could encode the mechanisms that drive or constrain these correlations.

This "middle ground" suggests that while Bell's theorem showcases the non-classical correlations of quantum mechanics, it may not definitively prohibit any and all local hidden variable frameworks. Instead, it bans only those that are too narrowly defined, those lacking additional probabilistic rules or extra-dimensional channels to capture the complexities of quantum behaviour. The wave function could be an emergent mathematical tool summarizing all possible states, while a deeper, more local structure underlies it, hidden by both our experimental constraints and our limited viewpoint in conventional space-time.

Bell's theorem and its experimental tests undeniably revolutionized our understanding of quantum mechanics, compelling physicists to grapple with the reality of non-local correlations. However, declaring them as absolute, final disproofs of local hidden variables may overlook important theoretical and experimental subtleties. The possibility of superdeterminism, reliance on specific experimental loopholes, and the potential for higher-dimensional explanations in theories such as string theory all point to a more nuanced conclusion.

Far from being a settled question, the nature of quantum entanglement and its compatibility with local mechanisms remains an open area of research, ripe for bold, out-of-the-box thinking. A hybrid view that marries the wave function's non-locality with local hidden variables (nestled in probabilistic or higher-dimensional frameworks) could offer a profound and consistent understanding of reality. In this spirit, continuing to probe the limits of quantum experiments, challenging conventional assumptions, and exploring richer theoretical landscapes may lead us to a more complete picture of our quantum universe.

CHPLR Approach

This is why I developed a new mathematical framework: Contextual Higher-Dimensional Probabilistic Local Realism (CHPLR). This theory provides a fresh perspective, proposing that phenomena appearing non-local within our familiar 3+1 space-time could be entirely local when viewed from a higher-dimensional framework.

CHPLR introduces a single, coherent structure that preserves a deeper form of locality while reproducing the characteristic non-local correlations observed in quantum mechanics. By embedding our 3+1 universe into a higher-dimensional manifold, incorporating contextual hidden variables, and relaxing the assumption of strict measurement independence, CHPLR reconciles Bell-inequality violations with a redefined notion of local realism, operating within this higher-dimensional context.

Ultimately, CHPLR serves as a comprehensive and testable theory, challenging the prevailing notion that quantum entanglement necessitates fundamental non-locality. Instead, it posits that quantum correlations arise from the projection of a fully local, higher-dimensional reality onto the constrained lens of our 3+1-dimensional experience. This framework opens the door to deeper explorations of the true nature of reality, bridging the gap between locality and quantum phenomena.

Objectives:

- **Preserve Local Causality in a Deeper Sense:** Show that all fundamental processes may be strictly local in a higher-dimensional manifold, even though, from a 3+1 perspective, particle correlations look "spooky."
- **Incorporate a Contextual Hidden Variable Structure:** Define hidden variables that adapt to measurement setups in a consistent, higher-dimensional manner, thus reproducing quantum correlations.
- **Explain Apparent Bell Violations:** Demonstrate that Bell test outcomes can be accounted for by relaxing certain standard assumptions (e.g., strict measurement independence) and introducing higher-dimensional locality.
- **Provide a Testable Framework:** Offer theoretical avenues for experimental validation or falsification, highlighting new predictions and subtle deviations that could be probed with next-generation technology.

Foundational Postulates:

Higher-Dimensional Space-time Manifold $M_{D>4}$

- Our observable 3+1 world is embedded in or emerges from a larger manifold with **D>4** dimensions.
- All fundamental fields, particles, and causal influences are governed by local laws in $M_{D>4}$, with no superluminal propagation in that higher-dimensional space.

Contextual Local Hidden Variables λ

- Each particle or quantum system carries a set of hidden variables $\lambda \in \Lambda$.
- These variables are *contextual*, meaning λ can adapt or branch in response to local conditions in $M_{D>4}$.
- The observed outcomes in 3+1 space-time are projections λ_{proj} of the underlying higher-dimensional λ.

Relaxed Measurement Independence

- We allow partial statistical correlation between λ and the experimental choices (e.g., measurement settings A , B).
- This correlation remains a local process within $M_{D>4}$, but can manifest in 3+1 as apparent "superdeterminism" or "fine-tuning."

Emergent Wave Function

- The quantum wave function ψ is an effective description arising from coarse-graining or integrating out the detailed local variables in $M_{D>4}$.
- Non-local correlations in ψ reflect higher-dimensional geometric constraints, *not* real-time superluminal interactions.

Mathematical framework:

Fields and Locality in $M_{D>4}$

- Let $\Phi(\mathbf{x})$ be a field defined on $M_{D>4}$ where **x** comprises coordinates beyond 3+1.
- Fundamental dynamics obey a local Lagrangian density $L(\Phi, \nabla_D \Phi)$ prohibiting causal signals from exceeding some speed C_D in **D**-dimensional space.
- A projection operator Π maps events $X \in M_{D>4}$ down to $((t, \vec{r}) \in M_{3+1})$

Contextual Hidden Variable Distribution

- Define a probability distribution for hidden variables:

$$p\,(\lambda, x)\ \ with\ \ \lambda \in \Lambda, x \in M_{D>4}.$$

- When a measurement setting **(A, B)** is chosen in 3+1, it corresponds to some local region or boundary condition in $M_{D>4}$. The relevant portion of **λ** is "projected" to λ_{proj}.
- The measurement outcome is given by a local function in higher dimensions,

$$O(A, B) = f\,(\lambda_{proj}, A, B)$$

Ensuring that, in the full manifold, no superluminal signalling is required.

Relaxed Measurement Independence

- Replace the strict factorization assumption $p(\lambda, A, B) = p\,(\lambda)\,p\,(A, B)$ with: $p\,(\lambda \mid A,\ B) = p\,(\lambda)\,[1 + \delta\,(\lambda,\ A,\ B)]$.
- **δ (λ, A, B)** encodes partial or context-sensitive correlations that remain local in $M_{D>4}$ but look "fine-tuned" in 3+1.
- By adjusting **δ**, one can systematically match the degree of quantum-like correlations observed in Bell tests.

Emergence of the Wave Function

- Propose a functional

$$\psi\,(\vec{r}_1, \dots, \vec{r}_n, t) = \int_{M_{D>4}} G\,[\Phi\,(x), \lambda\,]\,d\mu\,(x)\,,$$

That integrates or coarse-grains higher-dimensional variables into an effective amplitude **ψ**.

- The Born rule, $|\psi^2|$, naturally arises as a macroscopic probability measure of outcomes once hidden variables are averaged out.

Recovery of Quantum Phenomena

Violation of Bell/CHSH Inequalities

- The classical Bell or CHSH bound (e.g., **S ≤ 2**) is derived under assumptions that do not hold in CHPLR (namely, full measurement independence and a strict 3+1 classical probability space).
- In CHPLR, the generalized correlation function

$$E_{CHPLR}(A, B) = \langle f\ (\lambda_{proj}, A, B) \rangle$$

Depends on **δ (λ, A, B)** and higher-dimensional geometry. Tuning these elements can replicate quantum violations up to $2\sqrt{2}$, consistent with experimental data.

Reproduction of Standard Quantum Results

- Under minimal correlations **δ ≈ 0** (i.e., approximate "free will" in choosing settings) and typical detector efficiencies, CHPLR's predictions align with the standard quantum mechanical formalism.
- This ensures that most conventional experiments, which assume random measurement choice and effectively high detection efficiency, see no immediate deviation from standard quantum results.

Potential Deviations

- In extreme high-precision tests or under artificially contrived measurement-setting correlations, CHPLR may yield small but detectable deviations from standard quantum theory.
- Such anomalies, if observed, would support the existence of a deeper higher-dimensional local structure that underpins entanglement-like phenomena.

Experimental Implications

Loophole-Free Bell Tests Revisited:

- Even "loophole-free" Bell experiments might not fully close the door on hidden correlations arising from higher-dimensional local processes.
- Novel tests involving ultrafast random-setting generators, cosmic photon-based random seeds, or advanced space-like separation techniques could challenge CHPLR to ensure **δ (λ, A, B)** is genuinely negligible.

Probes of Higher Dimensions:

- If extra dimensions exist, high-energy physics or cosmological observations might reveal subtle signatures (e.g., Kaluza-Klein resonances, gravitational wave anomalies).
- These signatures, if confirmed, lend plausibility to the central premise that fundamental locality is upheld in $M_{D>4}$

Contextual Variation Experiments:

- Systematically altering measurement contexts or introducing dynamic feedback loops between detected outcomes and setting choices could expose small contextual dependencies predicted by **δ (λ, A, B).**
- Any confirmed correlation between measurement settings and particle states, beyond standard quantum predictions, would be a hallmark of CHPLR.

Ultimately, quantum entanglement suggests that our reality operates according to principles more akin to a computational system than a purely physical one, where higher-dimensional logic and pre-encoded rules govern the behaviour of particles and their interactions. This perspective invites further exploration into the fundamental nature of reality, potentially bridging physics and computational science in profound ways.

Location-Based Determinism Theory

Looking at the universe and how it works, I developed this theory to explain a curious paradox: how can a universe that appears random on the surface be governed by deterministic laws underneath? Consider a computer program that generates "random" numbers. In truth, these numbers are not random; they are produced by a deterministic algorithm that depends on an initial "seed." If you run the same program with the same seed, you get the same output. Now, imagine that the entire universe behaves in an analogous manner: there is some sort of fundamental grid, a cosmic checkerboard of infinitesimal squares or "dots", and the outcome of any given event is determined precisely by its location on this grid, combined with the local forces acting on it.

This leads to a radical but tantalizing hypothesis: nothing is truly random; we perceive randomness only because we rarely (if ever) occupy the exact same location in the grid with the exact same local conditions more than once. In principle, if the universe could return to exactly the same configuration (including the same forces such as gravity, electromagnetic fields, etc.) and we repeated an experiment at that same quantum-level "dot," the outcome would be identical.

I call this the Location-Based Determinism (LBD) Theory. It proposes that when we talk about determinism, we are referring to the interactions between fundamental particles and forces. The cosmic grid, composed of these infinitesimal points or "dots", acts as a framework that governs the range of possibilities available at any given moment. For example, as the Earth moves through space along these grids, they provide a parameter of possibilities for everything on it. This means that when you move your hand in three-dimensional space, that motion is one of the possibilities allowed by the interaction between your particles, the Earth's particles, and the forces acting on the grid.

However, you cannot move your hand into a four-dimensional direction or suddenly develop a four-dimensional mind, because such possibilities do not arise from the interaction between particles and the grid. The grid, combined with the deterministic behaviour of particles and forces, constrains what is possible and defines the rules within which reality unfolds.

While LBD Theory preserves a deep-rooted determinism, it also places it permanently out of reach of experimental replication, owing to cosmic motion and expansion. In doing so, it challenges the conventional view of quantum probability, free will, and the fabric of reality.

It implies that our perception of randomness stems from our inability to track every interaction between particles and the underlying grid in real-time.

Despite its radical nature, LBD Theory serves an important philosophical role: it underscores the tension between determinism and randomness, highlighting the subtlety of scale and the inescapable fact that our vantage point on Earth, hurtling through space-time, may never let us anchor experiments to a single "dot" on the cosmic grid. Yet, it invites us to reconsider the very foundation of reality. It suggests that while we live within a deterministic framework, our understanding of it is shaped by our position in space-time and the limits imposed by the grid's interaction with particles and forces. The debate it provokes, and the imaginative expansions of physical theory it may inspire, are perhaps reason enough to give it a thoughtful, if daring, consideration.

Key Tenets and Rationale

- **Quantized Cosmic Grid**
  The universe is hypothesized to be under-laid by a "grid" of infinitesimally small points, each with a predefined "seed value" that determines the outcome of events happening exactly there.
- **Influence of Local Forces**
  The outcome at each grid point is not static; it depends on local forces (gravity, electromagnetic fields, quantum fields, etc.). Thus, any location on the grid is associated with a family of outcomes, each triggered by a distinct configuration of local forces.
- **Motion and Expansion**
  Because the Earth (and everything else) is in constant motion and the universe is expanding, it is effectively impossible to reoccupy the exact same grid point under the exact same conditions. This perpetual motion creates the illusion of genuine randomness, because we never see the precise replication of an event in the same location and with the same force configuration.
- **Apparent Probabilities**
  In quantum mechanics, experiments yield probabilistic outcomes. Under LBD Theory, these probabilities reflect our ignorance of the underlying cosmic grid and the impossibility of isolating a single grid dot's contribution. Instead, we sample many grid dots at once, each with slightly different "seed values." Hence, we measure superpositions of those deterministic seeds and perceive "probability" where, at a finer scale, there is none.
- **Echoes of Determinism**
  In classical physics, determinism suggested that if we knew the precise initial conditions of every particle and field, we could predict the future. Quantum mechanics challenged this view, introducing fundamental randomness (or at least unpredictability). The LBD Theory attempts a bold reconciliation: the universe is still deterministic, but at such an unfathomably fine-grained level (the cosmic grid) that effectively no experiment can single out one dot from the next.
- **Relation to the "Cosmic Seed" Hypothesis**
  Just as pseudo-randomness in a computer is traceable to a seed, LBD Theory proposes that each dot in the grid is a seed. But each seed is not just a single number; it is an entire catalog of local outcomes as a function of local forces. The net effect is a labyrinthine puzzle of possibilities. Our instruments, no matter how refined, move across myriad grid points, painting the illusion of random distribution.
- **Metaphysical Implications**

If LBD Theory were true, "free will" and "chance" might be illusions, or at least phenomena emergent from our inability to isolate the cosmic grid's single-point influences. Determinism is pushed beneath the Planck scale and integrated with cosmic expansion and motion, reshaping existential questions about predictability, meaning, and the nature of reality.

Hypothetical (but Impossibly Difficult) Experiment

Requirements

- A region of space that is absolutely motionless relative to the cosmic grid. (Given universal expansion and gravitational perturbations, we currently see this as impossible.)
- An identical repetition of local forces, gravitational, electromagnetic, and so forth.
- The ability to confine an experiment strictly to a single point of the grid or, at worst, within a region that remains smaller than a single "dot."
- **Predicted Result**
  If these conditions could be met, rolling 10 dice (or performing any random-like quantum event) would yield the exact same result every time. The "dice" outcomes would be locked by the seed of that particular grid point, contradicting our everyday notion of randomness.
- **Practically Impossible**
  Because our solar system is moving at hundreds of kilometres per second around the galactic centre, the galaxy is moving relative to cosmic filaments, and the entire universe is expanding, there is no known way to stand "still" on one cosmic grid point long enough, under exactly the same conditions, to test the theory.

Although the experiment described above is, for all intents and purposes, impossible, we can still propose a mathematical model to capture the plausibility of LBD Theory:

Define the Grid

Let us define a continuous coordinate system $\boldsymbol{r} \in \mathbf{R}^3$ ($\boldsymbol{or\ R^4\ if\ we\ include\ time}$) that is overlaid with a discrete set of grid points $\boldsymbol{G} = \{\boldsymbol{g_i}\}$. We imagine $\| \mathbf{gi+1-gi} \| \approx \boldsymbol{\ell}$, where $\boldsymbol{\ell}$ might be on the order of (or even smaller than) the Planck length ($\approx \mathbf{10^{-35}}\ \boldsymbol{m}$).

Seed function

We introduce a seed function $\boldsymbol{S(g_i, F)}$ that describes the deterministic outcome at grid point $\boldsymbol{g_i}$ given a set of local forces $\boldsymbol{F}$. For example, for rolling 10 dice, let

$$\boldsymbol{S(gi, F) = (\omega_1, \omega_2, \dots, \omega_{10})},$$

Where each $\boldsymbol{\omega_i}$ is a specific integer from 1 to 6, determined by the seed. In principle, $\boldsymbol{F}$ includes gravity, EM fields, temperature, etc.

Event Probability

In real experiments, we sample not just one $\boldsymbol{g_i}$ but rather a region $\boldsymbol{R} \subset R^3$ containing many grid points. Each point $\boldsymbol{g_i} \in \boldsymbol{R}$ might see slightly different forces $\boldsymbol{F_i}$. Hence, the observed outcome distribution is

$$P(\omega) = |\frac{1}{R}| \sum_{g_i \in R} \delta(S(g_i, F_i), \omega),$$

Where δ is a Kronecker delta that is 1 if $\boldsymbol{S}$ (g_i, F_i) equals ω, and 0 otherwise, and |R| is the number of grid points in R.

Experimental Feasibility Parameter

Define a function

$$E(v, \Delta t, \nabla F)$$

That captures how fast the experiment's apparatus is moving ***(v)*** over a time interval $\boldsymbol{\Delta t}$, and how quickly the forces change in space ***(∇F)***. The smaller $\boldsymbol{v}$ and $\boldsymbol{\nabla F}$ are, and the shorter the $\boldsymbol{\Delta t}$, the closer the experiment is to sampling a single grid point with a single force configuration. Perfect stillness and no force gradient ($\boldsymbol{\nabla F} = \boldsymbol{0}$) would yield $\boldsymbol{E \to 0}$, which is the ideal scenario for LBD Theory validation.

Truth-Value Likelihood

We can propose a dimensionless "Truth-Value Likelihood" **Λ** for LBD Theory, given an experimental dataset **D**:

$$\Lambda(D \mid LBD) = exp[-\chi^2(D, model)],$$

Where χ^2 (D, model) measures the deviation of observed outcomes from the predictions of LBD-based sampling. One might interpret a higher $\boldsymbol{\Lambda}$ as lending more support to the theory, though under present conditions, LBD predictions typically align with standard quantum-mechanical probability, making $\boldsymbol{\Lambda}$ difficult to distinguish from conventional interpretations.

Quantum Tunnelling: The Magic of the Impossible

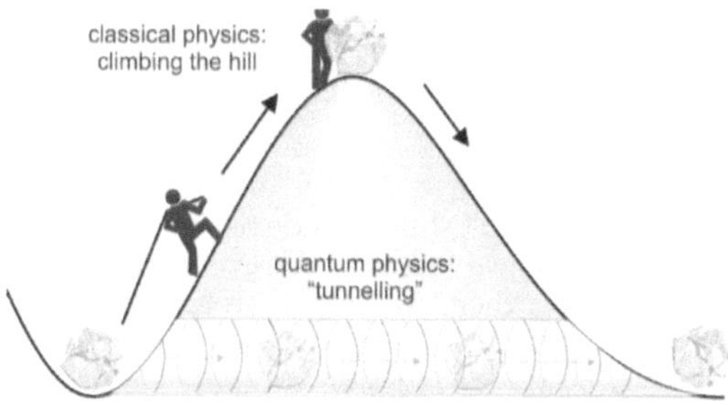

Imagine standing in front of a high wall. In the everyday world, if you don't have enough strength or energy, you can't climb over it or break through it. Now imagine that, somehow, you find yourself on the other side of the wall, even though you didn't climb, break, or pass through it in any obvious way. This might sound like magic or a glitch in reality, but in the quantum world, this phenomenon is real. It's called quantum tunnelling.

As discussed particles in the quantum world, like electrons, are not like tiny balls bouncing around; they also behave like waves. These waves represent the particle's possible locations and energies. When a particle encounters a barrier, like a hill it doesn't have enough energy to climb, classical physics says it must stop or bounce back. But because the particle is also a wave, part of that wave can "leak" through the barrier to the other side, even though it doesn't have enough energy to "climb" over.

Think of the particle as a ghost trying to cross a locked door. In classical physics, the ghost would have to break the door down to get through. But in quantum mechanics, the ghost is a bit "fuzzy" and spread out like a cloud. This fuzziness allows a tiny part of the ghost to appear on the other side of the door without breaking it, this is tunnelling.

Scientists have observed quantum tunnelling in real-world experiments. One of the most famous examples is the scanning tunnelling microscope (STM), a powerful tool that allows us to "see" individual atoms on a surface. This works because electrons tunnel through a tiny gap between the microscope's tip and the surface, creating a measurable current. The size of this current depends on the distance between the tip and the surface, directly demonstrating quantum tunnelling in action.

Quantum tunnelling is also well-described mathematically using the Schrödinger Equation. The probability of a particle tunnelling through a barrier can be calculated using this formula:

$$T \propto e^{-2ka}$$

Where:

k = $\sqrt{\frac{2m(V-E)}{\hbar^2}}$,

T = tunnelling probability,

m = particle's mass,

V = barrier height,

E = particle's energy,

a = width of the barrier.

This equation shows that the likelihood of tunnelling decreases as the barrier gets wider or taller. However, as long as the barrier isn't infinite, there's always a small chance for tunnelling to occur.

Quantum tunnelling might seem like a strange quirk of physics, but it's incredibly important. Without it, the sun wouldn't shine. Inside the sun, atomic nuclei fuse together to produce energy, but they don't have enough energy to overcome the repulsive forces between them. Tunnelling allows them to "leap" through this energy barrier and fuse, powering the sun and giving us life.

Quantum tunnelling exemplifies how quantum mechanics challenges classical intuition by revealing a probabilistic reality where even "impossible" events have calculable likelihoods. If the universe is a simulation, tunnelling might reflect computational optimization, enabling state transitions without expending the energy to overcome barriers, much like shortcuts in computational models.

This phenomenon ensures that improbable but critical transitions, such as nuclear fusion, remain possible, aligning with the idea of a universe designed to accommodate essential possibilities. By preserving computational or energetic resources while maintaining physical plausibility, tunnelling could be a mechanism in a simulated universe for achieving efficient resource use. Its probabilities, dictated by the wave function, suggest a system designed with probabilistic algorithms to ensure consistent yet dynamic outcomes.

In a finely tuned universe, whether natural or artificial tunnelling provides the flexibility for life and complexity to emerge under restrictive energy conditions, ensuring predictability within a probabilistic framework while enabling the existence and evolution of complex systems like stars and biological molecules. It highlights how reality systematically accommodates the unexpected, whether through natural laws or the optimized logic of a simulation.

The Heisenberg Uncertainty Principle (1927)

Werner Heisenberg, a young German physicist, was at the heart of the quantum revolution. By the mid-1920s, he was grappling with a central problem: how do we describe the behaviour of electrons within an atom? Niels Bohr had proposed that electrons occupied specific energy levels around the nucleus, but their precise behaviour was still cloaked in mystery. Traditional physics, with its insistence on exact positions and trajectories, seemed ill-suited for the quantum realm.

Heisenberg's breakthrough came in 1927. To appreciate its depth, imagine trying to measure a tiny, jittery insect darting around a dark room. You could use a flashlight to locate it, but the very act of shining the light disturbs the insect, changing its motion. Similarly, Heisenberg realized that measuring certain properties of a quantum particle, like an electron, inevitably disturbs it. This is not a matter of experimental limitation or technological shortcomings; it's a fundamental aspect of nature.

Here's where the mathematics comes into play. Heisenberg discovered that there's an inherent trade-off in our ability to measure two related properties of a particle. Specifically, the more precisely we measure a particle's position (where it is), the less precisely we can know its momentum (how fast and in what direction it's moving), and vice versa.

This relationship is encapsulated in a deceptively simple equation:

$$\Delta x \,.\, \Delta p \geq \frac{\hbar}{2}$$

The symbol **Δx** represents the uncertainty in position, **Δp** represents the uncertainty in momentum, and **ħ** (h-bar) is a constant derived from Planck's constant. The equation tells us that the product of these uncertainties can never be smaller than a fixed, tiny value. In essence, the universe imposes a **limit** on how much we can know about the quantum world. But why?

The implications of the **Uncertainty Principle** were revolutionary. In classical physics, if we knew the initial conditions of a system, say, the position and velocity of a ball, we could predict its future behaviour with perfect accuracy. Heisenberg's discovery shattered this deterministic worldview. At the quantum level, the universe operates probabilistically. We can only speak of probabilities and likelihoods, not certainties.

Heisenberg's idea wasn't initially embraced by everyone. Einstein, for instance, famously quipped,

"God does not play dice with the universe,"

Expressing his discomfort with the randomness at the heart of quantum mechanics. But as experiments accumulated, the Uncertainty Principle proved to be an undeniable cornerstone of quantum theory.

One fascinating aspect of the principle is how it challenges our intuition. In our everyday lives, we're used to the macroscopic world where uncertainty is negligible. You can measure the speed of a car and its location simultaneously with incredible precision. But at the atomic scale, the rules change, and Heisenberg's principle reigns supreme. It's a bit like discovering that the rules of a board game you've played your entire life are entirely different when you use a magnifying glass to examine the game pieces.

Heisenberg's Uncertainty Principle also opened the door to deeper philosophical questions. What does it mean to "know" something if there are limits to our knowledge? Does the act of observation create reality?

Today, the Uncertainty Principle isn't just a theoretical curiosity. It underpins technologies like quantum computing and quantum cryptography. It's a reminder that the universe is not a rigid machine but a dynamic, interconnected web of probabilities and possibilities.

In the context of a simulation, the uncertainty principle could function as a computational safeguard, strategically limiting the precision with which certain properties can be known. By restricting exactness, the simulation avoids the impracticality of storing or processing infinite data for every particle. Instead, it operates probabilistically, calculating only probabilities rather than deterministic states, thereby conserving computational resources and enhancing efficiency.

This probabilistic framework introduces an element of variability and complexity, essential for fostering emergent behaviours and dynamic interactions. By avoiding deterministic rigidity, the system achieves a balance, adhering to foundational rules while maintaining the richness and unpredictability necessary for a vibrant, interactive reality.

Heisenberg's principle, which limits our ability to precisely observe and measure reality, could be viewed in a simulated context as a boundary imposed by the system's programming. Similar to how video games restrict what players can access or interact with in real time, this principle may ensure that the quantum world remains partially unresolved, preserving the illusion of continuity while minimizing computational strain.

This raises the profound idea that observation itself could act as a trigger for rendering or updating the simulation. Until observed, particles remain in a state of superposition, allowing the system to defer computational effort until the exact moment it becomes necessary, akin to "lazy loading" in software development. Such efficiency ensures that resources are allocated dynamically, only when needed, maintaining a functional and responsive universe.

Furthermore, Heisenberg's principle provokes a philosophical question: could a self-organizing, organic system independently impose such constraints upon itself? Or are these limits more consistent with the hallmarks of an intelligently designed framework, where the boundaries are purposefully encoded to obscure the system's deeper mechanisms? If the latter, this design may reflect a desire to prevent full understanding of the program's structure, perhaps to deter replication or avoid acknowledgment of the foundational system.

Ultimately, the uncertainty principle portrays a universe governed by precise, probabilistic rules that balance knowledge and resolution. Its constraints suggest a system optimized for efficiency, where observation plays a pivotal role in shaping reality. This interplay of probabilistic outcomes, wave-particle duality, and observer-dependent states parallels the features of advanced simulations, where the underlying rules ensure both complexity and coherence without overwhelming the system.

By setting these limits, the uncertainty principle not only describes the behaviour of quantum mechanics but also hints at the possibility of a programmed reality. This system, designed to be dynamic, interactive, and computationally efficient, may reflect the hallmarks of a simulation where every detail serves a purpose, ensuring adaptability while shielding its origins from complete understanding.

Dirac Equation (1928) - Quantum Field Theory

In the annals of scientific history, few moments shine as brightly as the year 1928, when the brilliant physicist Paul Dirac introduced his eponymous equation. It was a revelation that bridged quantum mechanics and special relativity, reshaping our understanding of the universe and paving the way for quantum field theory. The story of the Dirac Equation is not just one of mathematical genius but also of a profound leap in conceptual thinking.

By the late 1920s, physics was at a crossroads. On one hand, quantum mechanics, with its probabilistic description of particles, was revolutionizing our understanding of the atomic world. On the other hand, Einstein's theory of special relativity had fundamentally altered our conception of space and time, showing that they were interwoven into a four-dimensional fabric. The problem? These two ground-breaking theories didn't play well together. Quantum mechanics worked beautifully for slow-moving particles but broke down when particles approached the speed of light. Special relativity, meanwhile, was indispensable for understanding high-speed phenomena but lacked a quantum description.

Physicists yearned for a unified theory that could reconcile these two pillars of modern science.

Paul Adrien Maurice Dirac was a man of few words but immense intellect. Born in 1902 in Bristol, England, Dirac possessed an extraordinary talent for mathematics. His approach to physics was driven by a belief in the power of mathematical beauty; he often argued that the elegance of an equation was a sign of its truth.

Dirac's quest began with a deceptively simple question: how could one describe the behaviour of an electron moving at relativistic speeds? *Schrödinger's* equation was inadequate because it didn't account for relativistic effects. Dirac needed a new equation, one that respected the principles of quantum mechanics and special relativity simultaneously.

The key lay in an earlier idea called the Klein-Gordon equation, which was an attempt to combine quantum mechanics and relativity. However, this equation had a critical flaw: it couldn't properly describe the electron's spin, an intrinsic quantum property. Spin is a bit like the way a spinning top has angular momentum, but in quantum mechanics, it's more abstract and doesn't correspond to actual physical spinning.

Dirac's breakthrough came when he realized that the problem required a new type of mathematical structure. Instead of using scalar wave functions (simple numbers), Dirac introduced something called ***spinors***. A *spinor* is a mathematical object with multiple components, much like a vector, but it transforms differently under rotations in space and time. By working with spinors, Dirac was able to incorporate the electron's spin naturally. The Dirac Equation can be written as:

$$\left(i\gamma^{\mu}\partial_{\mu} - m\right)\psi = 0$$

It describes the behaviour of relativistic particles like electrons, accounting for both their wave-like and particle-like nature. The term $\mathbf{i\gamma^{\mu}}$ incorporates the imaginary unit **(i)** and gamma matrices $\boldsymbol{\gamma^{\mu}}$ which are mathematical constructs ensuring the equation respects the principles of special relativity. These gamma matrices encode the relativistic behaviour of particles and allow the equation to address the complexities of space-time.

The derivative term, $\boldsymbol{\partial_{\mu}}$ represents changes in the wave-function $\boldsymbol{\psi}$ across both space and time. This wave-function is a key concept in quantum mechanics, containing all the information about the quantum state of the particle, such as its position and momentum probabilities. The **(m)** in the equation denotes the mass of the particle, connecting to Einstein's famous equation $\boldsymbol{E = mc^2}$, which relates mass and energy.

Together, these terms describe how the particle's quantum state evolves under the influence of space-time and mass. At its heart, the Dirac Equation describes how an electron's wave function evolves over time and interacts with the space-time around it. What made it revolutionary was its ability to predict phenomena that had never been observed before.

One of the most startling predictions of the Dirac Equation was the existence of antimatter. When Dirac solved his equation, he found solutions that corresponded to electrons with negative energy. Initially, this seemed nonsensical, how could energy be negative? But Dirac proposed a radical idea: these negative-energy solutions represented particles with the same mass as the electron but opposite charge. These particles, later named **positrons**, were discovered experimentally in 1932, confirming Dirac's prediction.

Antimatter

Antimatter can be thought of as the mirror opposite of ordinary matter. Every particle in the universe, from electrons to protons, has an antimatter counterpart with the same mass but opposite charge and quantum properties. For instance, the antimatter counterpart of the electron is the positron, which has a positive charge. When matter and antimatter meet, they annihilate each other, converting their entire mass into pure energy according to Einstein's famous equation, $\boldsymbol{E = mc^2}$.

The notion of antimatter being particles moving backward in time stems from their mathematical descriptions in quantum field theory. To better understand this, let's imagine time as a flowing river. Ordinary particles, like electrons, are like boats moving downstream. Antimatter particles, such as positrons, behave mathematically as if they're moving upstream, backward in time. However, from our perspective in the forward flow of time, antimatter still appears as ordinary particles but with opposite charges and properties.

This idea, though abstract, provides an elegant symmetry in physics. It doesn't mean that antimatter actually travels to the past in a practical sense. Instead, it's a way of describing how antimatter behaves within the framework of the universe's fundamental laws. This perspective challenges our intuitive understanding of time and causality and suggests that antimatter might hold the key to unlocking deeper truths about the nature of time itself.

Baryon Asymmetry

The existence of antimatter is rooted in the fundamental symmetries of nature. When the universe was in its infancy, it was an unimaginably hot and dense soup of energy. According to quantum mechanics, high-energy photons could spontaneously convert into particle-antiparticle pairs. These pairs would typically annihilate each other almost instantly, reverting back into energy. However, for reasons we do not yet fully understand, this process slightly favoured matter over antimatter. For every billion particle-antiparticle pairs, one extra particle of matter was created. This tiny imbalance, a result of mechanisms like CP (charge-parity) violation, is what allowed the matter we observe today to survive and form stars, galaxies, and eventually us.

In the first fractions of a second after the Big Bang, the universe was a chaotic dance of matter and antimatter, constantly being created and destroyed. As the universe expanded and cooled, the energy levels dropped, halting the creation of new particles. At this point, any remaining matter and antimatter would annihilate. If the universe had started with equal amounts of both, all matter would have been destroyed, leaving behind a universe filled only with light, a desolate void without planets, stars, or life.

However, a small asymmetry in the laws of physics allowed some matter to survive. This phenomenon, known as **symmetry breaking**, is still not fully understood but is crucial to the existence of the universe as we know it. The leftover matter from this imbalance makes up the galaxies, stars, and planets, and everything else, in our universe today.

Antimatter Universe

What if the asymmetry had gone the other way? It's theoretically possible that a universe could exist where antimatter dominates instead of matter. Such an antimatter universe might look identical to ours, with antimatter stars and planets. However, this leads to a deeper question rooted in the conservation of energy.

According to the first law of thermodynamics, energy cannot be destroyed; it can only transform. If antimatter existed in equal quantities to matter at the beginning of the universe, but it no longer appears readily in our observable universe, then logically, that antimatter must still exist somewhere. Antimatter is energy, and since energy cannot disappear, it raises the tantalizing possibility that antimatter resides in a separate domain, perhaps an antimatter universe.

This could be imagined like two balloons sitting next to each other, with their edges repelling one another to avoid mixing. The universes of matter and antimatter might coexist but remain distinct, separated by some cosmic boundary. This idea not only aligns with the conservation of energy but also provides a plausible framework for the existence of an antimatter-dominated universe. If such a universe exists, it would be inherently isolated from ours, as any contact between the two would result in annihilation.

Additionally, we live in a logic-based universe, not a magical one. Theoretical models suggest that the possibility of parallel universes exists. One logical way to create parallel universes without them mixing could be through this matter-antimatter divide. Antimatter's absence from our observable universe may serve as indirect evidence of such a parallel antimatter universe. This symmetry and separation would allow both universes to exist simultaneously without catastrophic interactions, further supporting the idea of a multiverse.

These parallel universes remain speculative but provide an intriguing avenue for thought and exploration. The fact that we cannot easily detect antimatter might simply mean it is part of a vast, inaccessible cosmic structure, a reflection of our own universe, yet fundamentally different.

Antimatter Bombs : Unlimited Energy

The annihilation of matter and antimatter releases energy with unmatched efficiency. An antimatter bomb, if it were feasible to create, would convert 100% of the mass of the matter and antimatter into energy. By contrast, a nuclear bomb relies on fission or fusion, which only converts a tiny fraction of the material's mass into energy. For instance, 1 gram of antimatter could produce an explosion equivalent to about 43 kilotons of TNT, far exceeding the destructive power of traditional nuclear weapons.

Moreover, antimatter offers the tantalizing possibility of creating nearly limitless energy. Unlike nuclear fusion or fission, which have by-products and require massive infrastructures, antimatter's annihilation with matter is a clean and direct process, producing pure energy. With sufficient advancements, antimatter could enable the construction of never-ending energy machines, removing the need for massive structures like

Dyson Spheres.[20]. Such technology could provide energy far beyond the current scope of human capability, transforming civilization on an unprecedented scale. However, producing

[20] A Dyson Sphere is a hypothetical megastructure proposed by physicist Freeman Dyson, designed to capture and harvest the energy output of a star. It envisions a shell or swarm of solar-collecting satellites surrounding the star to harness its immense energy for advanced civilizations.

and storing antimatter remains a monumental challenge, as even creating milligrams of antimatter today would require more energy than it could produce.

Imagine a cosmic bakery making cookies (particles). For every cookie made, an anti-cookie is also made. Normally, cookies and anti-cookies would cancel each other out, leaving an empty plate. Yet somehow, at the end of the process, there are a few cookies left on the plate while the anti-cookies are mysteriously missing. We're the leftover cookies, and the mystery of the missing anti-cookies is what physicists are trying to solve.

Antimatter is not the only mechanism that, in theory, could unlock colossal or even "limitless" sources of energy. Several other hypothetical processes have been proposed as ways to harness staggeringly large amounts of power, potentially enough to simulate entire universes.

Kugelblitz Energy

A "*Kugelblitz*," in theoretical physics, is a black hole formed not from collapsing matter but from an intense concentration of electromagnetic radiation (photons). If enough energy is focused into a small enough region of space-time, that energy could theoretically curve space-time to the point of forming a black hole, this is called a *kugelblitz* (German for "ball lightning"). Instead of using stars' gravitational collapse, you'd be using beams of incredibly high-energy light to create a miniature black hole.

Once a kugelblitz (or any small black hole) forms, it should emit radiation over time through Hawking radiation. By carefully managing this miniature black hole, one could, in principle, "feed" it matter or energy and harvest the resulting Hawking radiation as a continuous power source. The smaller the black hole, the more intense its Hawking radiation, at least until it evaporates entirely. In an advanced civilization's energy scheme, a *kugelblitz* reactor might become a self-sustaining power plant, so long as the black hole is kept stable and regularly fuelled.

- **Energy Requirement:** The amount of energy needed to create a black hole out of photons is astronomical. Generating such energy in a controlled manner surpasses our current engineering capabilities by many orders of magnitude.
- **Containment and Control:** Safely containing and managing a black hole, no matter how small, represents an unprecedented feat of engineering. Allowing a black hole to drift outside its containment area or letting it lose too much mass too quickly could lead to disastrous consequences (either uncontrolled evaporation bursts of radiation or a loss of the energy source altogether).

Despite these difficulties, the kugelblitz idea remains a fascinating glimpse into potentially civilization-redefining energy technologies.

Zero-Point Energy Extraction

Zero-point energy refers to the lowest possible energy that a quantum system may have, stemming from the inherent fluctuations of quantum fields, even in what we classically think of as "empty space." The vacuum isn't truly empty; it seethes with virtual particles popping in and out of existence. The Casimir effect is one tangible demonstration of zero-point energy: two uncharged metal plates brought very close together can measure a small attractive force due to changes in the quantum vacuum energy between them.

If one could find a way to systematically harness these quantum fluctuations, essentially extracting energy from "nothing", the supply might be as vast as the universe itself. The vacuum is theorized to contain immense energy density; if even a tiny fraction of this vacuum energy could be tapped, it would far surpass our conventional energy sources.

- **Fundamental Physics Limitations:** While zero-point energy is acknowledged to exist, mainstream physics does not currently have a proven mechanism for extracting it on a net-positive, industrial scale. Any attempt to do so bumps into paradoxes and issues related to the laws of thermodynamics and quantum field theory.
- **Engineering Hurdles:** Even if a small effect (like the Casimir force) can be demonstrated, scaling this to produce substantial amounts of usable power is another matter entirely. Our technology is nowhere near the point of making "vacuum energy generators."

Quark-Gluon Plasma

Quark-Gluon Plasma (QGP) is a state of matter believed to have existed just microseconds after the Big Bang. In this ultra-high-energy environment, quarks (the fundamental constituents of protons and neutrons) and gluons (the force carriers of the strong nuclear force) exist freely, not bound up into hadrons like protons and neutrons. Today, scientists recreate fleeting instances of QGP in particle accelerators like the Large Hadron Collider (LHC) by colliding heavy ions at near-light-speed.

If we could somehow sustain and harness QGP, it might provide a new realm of nuclear or sub-nuclear energy release. The strong force is the most powerful of the four fundamental forces, and direct manipulation of quarks and gluons could yield extraordinary energy outputs, potentially even more than conventional nuclear fusion.

- **Extreme Conditions:** Generating QGP requires temperatures and energies trillions of degrees hotter than the Sun's core, conditions that so far exist only momentarily in high-energy colliders.
- **Containment & Duration:** Even if we could reach these energies, keeping QGP stable for longer than a fraction of a second is an enormous challenge. Current technology only allows us to briefly study it before it cools back into ordinary matter.
- **Practical Utilization:** We're still at the stage of basic research. The step from ephemeral laboratory experiments to a controlled power source is nearly inconceivable with our present capabilities.

Still, the very existence and study of QGP show that there may be untapped forms of matter and energy beyond anything we have yet harnessed.

Vacuum Decay

Vacuum decay is a concept that arises from quantum field theory, suggesting that our universe might be in a "metastable" vacuum state rather than a perfectly stable one. In a metastable state, there is a possibility, albeit extraordinarily small, that a quantum tunnelling event could trigger the vacuum to transition to a lower energy state (the "true vacuum"). If this transition were to occur, it would expand outward at nearly the speed of light, effectively rewriting the laws of physics and destroying everything in the process.

If our current vacuum state does indeed sit higher on the energy scale than the true vacuum, then a transition would release an immense amount of energy, fundamentally altering the universe's structure. From a purely theoretical standpoint, one might imagine harnessing some fraction of that released energy before the bubble of true vacuum annihilates reality as we know it.

- **Universe-Ending Consequences:** Attempting to trigger vacuum decay or harness it is, by its very nature, an existential risk for the entire universe. It's not a localized event like a bomb going off; it's more akin to rewriting the cosmic rulebook.
- **Speculative Physics:** We don't know with absolute certainty whether our vacuum is metastable. Current data from particle physics (including the mass of the Higgs boson) suggests it might be "barely stable" or metastable on cosmological timescales, but that is far from a green light to exploit or even attempt to induce it.

Given the catastrophic outcome of an actual vacuum decay, it stands as perhaps the most extreme example of "energy release" in all of physics and is wisely relegated to theoretical musing rather than practical pursuit.

Nuclear fusion

Nuclear fusion is the process by which atomic nuclei combine to form a heavier nucleus, releasing an enormous amount of energy in the process. This is the same mechanism that powers stars, including our Sun, where hydrogen nuclei fuse into helium under immense pressure and temperature. Fusion releases energy because the mass of the resulting nucleus is slightly less than the sum of its parts, and this mass difference is converted into energy according to$E = mc^2$. The potential for harnessing nuclear fusion on Earth lies in its ability to provide a nearly limitless, clean energy source, as it produces no long-lived radioactive waste and relies on abundant fuel, such as isotopes of hydrogen (deuterium and tritium).

However, achieving and sustaining the extreme conditions required for fusion, millions of degrees of temperature and strong magnetic confinement, remains a formidable challenge. If mastered, nuclear fusion could revolutionize energy production, offering power on a scale far beyond current methods without contributing to climate change or environmental degradation.

Dirac's introduction of spinors and gamma matrices invites a profound question: how can such intricate mathematical constructs, perfectly suited to describe the quantum world, arise organically rather than being programmed? Spinors, with their multi-component nature, seamlessly encode properties like spin and relativistic transformations, while gamma matrices ensure consistency with special relativity, preserving the fabric of space-time. These constructs resemble advanced "data structures" in computational systems, optimized to model complexity with precision. Could such elegance emerge by chance, or does it point to the meticulous design of a simulation?

The implications grow deeper when we consider antimatter. The Dirac Equation predicts its existence with striking symmetry, raising the possibility of a multiverse where matter and antimatter are partitioned into separate "domains." In computational systems, isolating incompatible components ensures stability and prevents interference, a principle that seems mirrored in the cosmic separation of matter and antimatter. Is it plausible that this arrangement arose naturally, or does it reflect a deliberate effort to optimize the simulation's structure?

Annihilation, the process where matter and antimatter convert into pure energy, further underscores the parallels to computational design. This phenomenon functions like a reset mechanism, efficiently eliminating excess particles and maintaining balance within the system. The perfect conservation of mass and energy in this process echoes how computational systems clear memory to enhance performance. Can such efficiency be purely coincidental, or does it hint at a universe governed by programmed rules?

The Dirac Equation not only bridges quantum mechanics and relativity but also introduces concepts like spin, baryon asymmetry, and antimatter, features that contribute to the universe's stability and diversity. These elements suggest an underlying logic, as if the cosmos were calibrated with algorithms designed to maintain coherence and adaptability. How else can we explain the uncanny resemblance between the principles governing our universe and those that underpin computational systems?

Dirac's work reveals a universe where mathematical beauty and symmetry seem tailored to ensure a dynamic yet orderly reality. The existence of antimatter, its annihilation properties, and the possibility of parallel universes all align with principles of computational efficiency. Can we truly attribute such precise harmony to random chance, or does it reflect the hallmarks of a sophisticated simulation?

Ultimately, the question persists: how can the universe's mathematical elegance, optimization, and logical design arise without a programmer? Dirac's equations, and the phenomena they describe, challenge us to confront the possibility that our reality operates not by coincidence but through the deliberate coding of a system far beyond our comprehension.

The Big Bang (1949)

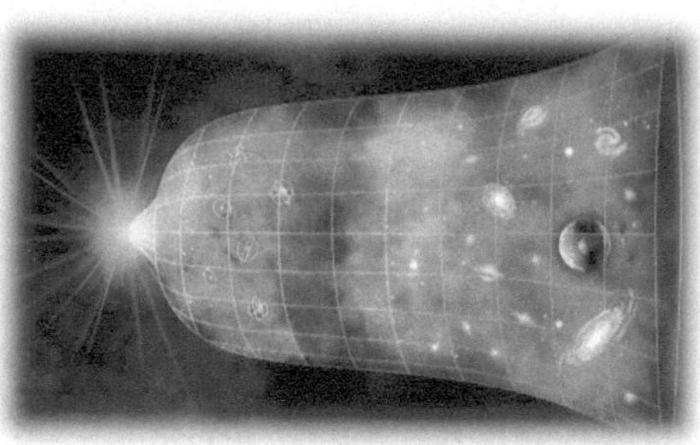

The Big Bang theory is one of the most significant scientific concepts that underpins our understanding of the universe's origins. It represents the prevailing cosmological model describing how the universe expanded from an extremely hot, dense state to its current form.

The term "Big Bang" was first coined in 1949 by the English astronomer Fred Hoyle, who ironically intended it as a derisive term for a theory he opposed. However, the idea that the universe had a finite beginning dates back earlier, rooted in both philosophical and observational traditions.

Building on Hubble's Law, George Gamow, along with his collaborators *Ralph Alpher* and Robert Herman, laid the theoretical foundation for the Big Bang in the 1940s. Their work was pivotal in predicting the existence of the Cosmic Microwave Background (CMB) radiation, a relic of the early universe.

George Gamow was a Russian-American physicist whose work spanned nuclear physics, quantum mechanics, and cosmology. He envisioned a universe where heavy elements were synthesized during a hot, dense beginning, an idea derived from his work on nuclear reactions.

Ralph Alpher, a student of Gamow, collaborated with him to write the seminal paper "The Origin of Chemical Elements" in 1948. This paper, often referred to as the *"alpha-beta-gamma"* paper (a playful nod to their names), outlined the processes of nucleosynthesis in the early universe.

Robert Herman, another collaborator, worked with Gamow and Alpher to refine the theoretical predictions about the conditions of the early universe. Together, they predicted the temperature of the CMB, a key piece of evidence later observed in 1965 by Arno Penzias and Robert Wilson, confirming the Big Bang theory.

The Big Bang theory posits that the universe began approximately 13.8 billion years ago from an initial singularity, a point of infinite density and temperature. This event marked the beginning of space-time itself, along with the emergence of matter, energy, and the physical laws that govern the cosmos. In this framework, time as we understand it also came into existence. Before the Big Bang, the concept of "before" loses its meaning because time itself did not exist. The Big Bang signifies the birth of time, space, and the universe's physical dimensions.

In the moments following the Big Bang, the universe underwent rapid expansion, known as cosmic inflation. During this phase, quantum fluctuations were stretched to macroscopic scales, seeding the large-scale structures we observe today, such as galaxies and clusters. As the universe expanded and cooled, it transitioned through several key epochs:

- **The Planck Epoch (0 *to* 10^{-43} seconds):** The earliest known period of the universe, where quantum gravity dominated, and our understanding of physics breaks down.
- **The Grand Unification Epoch (10^{-43} *to* 10^{-36} seconds):** Forces like electromagnetism, strong nuclear, and weak nuclear began to differentiate.
- **Cosmic Inflation (10^{-36} *to* 10^{-32} seconds):** The universe expanded exponentially, smoothing out irregularities.
- **The Quark Epoch (10^{-12} seconds):** Elementary particles such as quarks and gluons dominated.

- **The Nucleosynthesis Epoch (3 to 20 minutes):** Light elements like hydrogen, helium, and traces of lithium were formed.
- **Recombination (380,000 years):** Electrons and protons combined to form neutral atoms, allowing photons to travel freely, creating the CMB radiation.
- **The Formation of Structures:** Over billions of years, gravitational forces caused matter to coalesce into stars, galaxies, and clusters.

The Big Bang theory is supported by several lines of evidence. One major piece of evidence is the Cosmic Microwave Background (CMB), a faint glow of radiation predicted by Gamow, Alpher, and Herman and discovered by Penzias and Wilson. The CMB provides a snapshot of the universe when it was just 380,000 years old. Another key piece of evidence is Hubble's Law. Additionally, the abundance of light elements such as hydrogen, helium, and lithium closely aligns with predictions from Big Bang nucleosynthesis models. Lastly, the large-scale structure of the universe, including the distribution of galaxies and clusters, can be traced back to quantum fluctuations during cosmic inflation.

Flat Universe

Throughout history, our pursuit of a grand cosmological picture has hinged on what can seem like trifling details. Centuries ago, mathematicians spent untold effort wrestling with Euclid's fifth postulate, only to realise that alternative geometries, hyperbolic, spherical, or otherwise, might be just as mathematically consistent as the seemingly "obvious" flat case. This sudden freedom in geometry reverberates right through modern cosmology, where the fundamental question of the universe's curvature is anything but settled. Observations of the cosmic microwave background, supernova data, baryon acoustic oscillations, and the like converge, or occasionally clash, in painting a surprisingly nuanced portrait of the universe's shape. Should we trust appearances that the cosmos is flat, or does a slight curve in either direction lurk just beyond our present measurements?

When the Planck Collaboration released its 2018 legacy data, it appeared to bolster the long-held conviction that our cosmos is nearly flat, with its curvature parameter hovering so close to zero as to be indistinguishable from no curvature at all. Yet the subtleties of the Planck analysis revealed something more disquieting. In the paper "Planck evidence for a closed Universe and a possible crisis for cosmology", Di Valentino, Melchiorri, and Silk discuss an intriguing preference for positive curvature, suggesting that if one looks at certain Planck observations in isolation, particularly those relating to gravitational lensing of the cosmic microwave background, one obtains a fit indicating the universe could be closed (like a giant hypersphere), at odds with other data sets that point to near-flatness. This discrepancy ignited discussion of a possible cosmological "crisis," in which different observational windows yield contradictory impressions about whether space-time is truly flat or gently curved.

On one hand, the conceptual underpinnings of cosmic inflation offer a neat resolution to the "Why flat?" question. Inflation stretches any initial geometry, whether saddle-like or spherical, so dramatically that, at present, the cosmos would appear almost perfectly flat on the largest scales we can observe. Even a minute tilt away from zero curvature would, over billions of years, grow or diminish in a way that leaves behind an unequivocal imprint on the cosmic microwave background. Yet the data gleaned from Planck, which ideally should confirm or deny these theoretical assumptions, can sometimes resist neat classification. Combined analyses often prefer near-zero curvature, but subtle subsets of the data (like lensing reconstructions) hint at curvature that is small but non-zero. This conflict between lines of evidence not only complicates the standard cosmological model but also rekindles more radical speculations about the very nature of reality.

Perhaps, then, we are dealing with something beyond a simple question of flat versus curved. For centuries, geometry seemed merely a matter of mathematics, a logical structure describing space. But once we accept that, in theory, the universe could have been highly curved or entirely open, it feels almost contrived that we live in a place so finely balanced it hugs perfect flatness with near-infinite precision. This delicate cosmic balance has led many to question whether there is something deeply orchestrated at play: a cosmic "coding" that ensures not only the constants of nature fall into life-friendly niches, but that the entire stage is pegged to a geometry that just happens to be so pristine.

Some have posited this as evidence for a grand simulated universe, an environment programmed to present us with convincing physics and a geometry of zero curvature. Why else would it all line up so remarkably, if not by design? Alternatively, one might recall more exotic views, such as a holographic framework or the concept of Location-Based Determinism Theory, in which every local event is dictated by its position on a hidden cosmic grid. If the large-scale universe truly were flat in an absolute sense, one might well suspect it of being the ultimate engineered backdrop, designed so that physical processes everywhere obey identical, unerringly consistent rules.

And yet, from another angle, the apparent tension between Planck's closed-universe hint and the broader cosmological data set might itself demolish any illusions of a perfect geometry. If fresh observations or refined analyses confirm that the universe is indeed curved, ever so slightly, then the myth of an absolutely flat cosmos would unravel, puncturing the entire notion that inflation had pinned curvature at zero forevermore. On a philosophical level, even a tiny curvature emphasises that reality is not forcibly "elegant." If a bit of curvature is present, we can reimagine the cosmos as an unfolding tapestry, one that does not rely on improbable exactitude but on more intricate physical processes that still yield the illusions of near-flatness over local scales. After all, even on a planet's surface, a small region can appear perfectly flat to the observer standing upon it, whereas the global shape might be more richly contoured.

The heart of the matter is that cosmology today is driven not merely by theoretical aesthetics but by precise observational campaigns, from advanced surveys of the cosmic microwave background to exquisite measures of baryon acoustic oscillations. Paradoxically, the more precise these observations become, the more we edge towards "crises" where each new decimal place unsettles our best theories. The interplay between Planck's lensing signal and other data exemplifies this: perhaps we have uncovered a genuine cosmic feature, positive curvature in the geometry of the universe, or perhaps the tension owes to systematic effects or subtle statistical anomalies. Either conclusion would recalibrate what we assume about the standard model of cosmology, including its apparent near-flatness.

Meanwhile, from a philosophical perspective, consider the staggering complexity of forming a universe. If cosmic geometry were wholly "natural," we might expect it to meander through various curvatures. Yet, if observational results persist in indicating something so precariously balanced around zero, assuming future data do not tilt the scales towards a definitively curved outcome, then the question resurfaces with greater force: is such pinpoint flatness an accidental alignment, or an elaborate result of underlying design?

Enter the more radical line of thinking. If we cannot comfortably ground cosmic flatness in conventional physics, might it belong to a realm beyond standard space-time understanding, a sign that we dwell in an artificially constructed setting? The notion that we are embedded in an enormous simulation has shed much of its pulp-fiction aura in recent years, now discussed in serious scientific and philosophical contexts. In that light, a precisely flat universe becomes one of many curious "coincidences" that could be reinterpreted as evidence of design rather than chance.

We might pair that with the concept of a fundamental cosmic grid, akin to the suggestion that all events are pinned to location-based determinism. In such a scenario, the grid itself could enforce an overarching flatness, akin to a cosmic chessboard with each move determined by the square on which it occurs. Beneath the veneer of chaos, quantum fluctuations, cosmic expansion, and gravitational lumps, there might lurk a deterministic substrate that choreographs every outcome. If so, then the near-flatness of the observable universe would be no more than a by-product of the underlying grid's constraints, mandated from the start.

Yet for all these grand possibilities, it remains entirely plausible that the emergent story, once the dust settles, will reveal a modest but non-zero curvature, dissolving the entire puzzle of "Why is everything precisely flat?" in a single stroke. The closed-universe solution, hinted at by the Planck data, might dovetail better with fresh surveys, especially as we refine our grasp on late-universe expansions and gravitational lensing effects. If a consensus emerges for a marginally closed geometry, then cosmic flatness would occupy a brief epoch in scientific history, an approximation turned dogma before the evidence grew robust enough to supplant it.

Ultimately, we find ourselves suspended between elegant theoretical frameworks and empirical data sets that clash or converge in unexpected ways. If forthcoming experiments confirm a truly flat universe, then the notion of a contrived or programmed reality remains in play, some might say it gains momentum. Conversely, if future observations side with the present hints of curvature, we may discover that we never really dwelt on a knife-edge of perfect geometry at all. The universe might well be an asymmetrical expanse, complete with subtle arcs and dips, no more "flat" than the surface of a balloon is "square."

Either outcome, strict flatness or slight curvature, poses challenges to our current cosmological paradigm. Whether we are forced to admit that inflation has oversimplified reality or that all the secrets of the universe were coded into it from inception, we confront possibilities as monumental as they are unsettling. From ancient debates on parallel lines to Planck's precise but perplexing findings, the question of cosmic curvature cuts to the heart of who we are and how the cosmos truly operates.

And so we circle back to that single troublesome sentence in Euclid's timeless tome. The nature of parallel lines was once the pinnacle of mathematical conundrums. Now, that quandary reverberates in the darkest depths of space, where enormous triangles traced by primordial light waves might, if measured carefully, proclaim whether the universe is curved, flat, or something else entirely. In light of the difficulties and tensions chronicled in recent analyses, we are reminded that flatness is never merely a default state of being. Should it turn out the cosmos truly is and always has been flat, it would be a cosmic coincidence so extraordinary that one might well suspect an underlying design, a hidden algorithm, or a broader determinism embedded in the very fabric of reality.

Nothingness

If antienergy exists, nothingness exists.

The concept of "nothingness" has been a topic of intense philosophical and scientific inquiry, often defined as a state where absolutely nothing exists, no matter, energy, space, time, or even the laws of physics. However, the nature of our universe and its origins challenges the feasibility of such a state. Instead, the evidence suggests that true nothingness is impossible, as creation, existence, and the emergence of laws necessitate a foundational context or super-system.

Quantum fluctuations demonstrate that even in regions of apparent emptiness within our universe, activity persists. These spontaneous, temporary changes in energy within a vacuum, governed by the principles of quantum mechanics and the uncertainty principle, give rise to virtual particles, pairs of particles and antiparticles that momentarily appear and then annihilate each other.

This phenomenon illustrates that a vacuum, a region devoid of matter, is not 'nothing.' Instead, it is imbued with energy, governed by physical laws, and constantly buzzing with activity. This indicates that within the universe, true nothingness cannot exist, as even the simplest 'empty' space is defined and shaped by the laws of our cosmos.

Existence is evidence that nothingness cannot exist.

The Planck Epoch, the earliest known period of the universe, marks the point at which space-time, energy, and the physical laws we observe came into existence. If this epoch represents the beginning of everything, it raises profound questions about the singularity that preceded it. Where did this singularity exist? What laws allowed it to exist? If the singularity was a point of infinite density and temperature, its very existence implies a framework or super-system that governed its properties. Such a super-system could be another universe, a higher-dimensional space, or a multiversal framework with entirely different laws.

This relationship between systems and their super-systems is key. For a subsystem like our universe to exist, it must arise within or be connected to a super-system that establishes the conditions for its emergence. A subsystem cannot exist in isolation, as it would lack the foundational rules that define its properties and behaviours. The creation of our universe, complete with its finely tuned physical constants and coherent laws, suggests that it originated within a larger super-system that predated it. This super-system provided the meta-laws or conditions that allowed the singularity to transition into the expansive universe we observe today.

The multiverse hypothesis illustrates this idea well. In this model, individual universes are like "bubbles" within a larger multiversal "foam." Each universe operates as a subsystem, governed by its unique set of laws. The multiverse, as the super-system, establishes the probabilities and mechanisms by which these universes form. Even the "space" between universes, often described as void, cannot be true nothingness. It is governed by the meta-laws of the super-system, which allow for the emergence of these bubble universes.

This interdependence highlights a critical problem with the concept of nothingness. If our universe required a super-system to exist, then the absence of any system, a state of absolute nothingness, is logically impossible. The singularity could not exist without a framework. The transformation of the singularity into a universe with space, time, and energy required pre-existing conditions, whether these were meta-laws, higher dimensions, or a multiversal framework.

Moreover, if we consider the idea of nothingness as the absence of not only matter and energy but also potentiality, the question becomes even more paradoxical. The creation of anything, including our universe, necessitates some form of potential, rules, or framework. If true nothingness existed, then nothing could ever arise from it, as there would be no context or mechanism for creation. This makes true nothingness an abstract, rather than a practical, concept.

The creation of our universe, with its intricate laws, points to the existence of a super-system that governed its emergence. Whether this super-system is a multiverse, a higher-dimensional reality, or something entirely beyond our comprehension, it underscores the impossibility of true nothingness. Existence, it seems, is an intrinsic property of reality, and the idea of nothingness, a state without any existence, laws, or potential, is incompatible with everything we observe and understand about the universe.

The Standard Model of Particle Physics (1970s)

In the grand tapestry of physics, the Standard Model is one of humanity's most profound achievements, weaving together the fundamental particles and forces that shape our universe.

By the mid-20th century, physics had undergone two transformative revolutions. Classical laws, like Newton's mechanics and Maxwell's equations, described the macroscopic world, while quantum mechanics, anchored by Planck's equation, Schrödinger's wave mechanics, and Heisenberg's uncertainty principle, revolutionized the microscopic realm. Yet, a crucial puzzle remained: these theories governed different scales, and attempts to unify them revealed contradictions. Einstein's general relativity explained gravity but clashed with the probabilistic nature of quantum mechanics.

The Standard Model addressed part of this challenge, not by uniting gravity, but by providing a framework for the other three fundamental forces: electromagnetic, weak nuclear, and strong nuclear. It established a deeper understanding of the quantum world, connecting the smallest building blocks of nature.

Physicists long suspected that matter was composed of smaller entities. Experiments in the mid-20th century revealed a zoo of particles, electrons, protons, neutrons, and their peculiar cousins like muons and neutrinos. This overwhelming diversity hinted at a hidden simplicity.

In the 1960s, Murray Gell-Mann's quark model provided a breakthrough. Protons and neutrons, previously thought to be indivisible, were shown to consist of smaller particles called **quarks**. Gell-Mann identified six "flavours" of quarks, up, down, charm, strange, top, and bottom, terms reflecting their quantum properties rather than physical attributes.

Quarks are bound by the strong nuclear force, mediated by particles called gluons. Aptly named, gluons act as "glue," confining quarks within protons, neutrons, and other hadrons. This binding mechanism forms the foundation of the Standard Model's depiction of matter.

If quarks and leptons are the players, the forces of nature define the rules of the game. The Standard Model describes three fundamental forces, each mediated by specific particles known as gauge bosons:

- **Electromagnetic Force**: Responsible for light, electricity, and magnetism, this force is mediated by photons, massless particles that interact with charged particles.
- **Strong Nuclear Force**: Binding quarks together, this force is mediated by gluons and holds the atomic nucleus intact, overcoming the repulsion between positively charged protons.
- **Weak Nuclear Force**: Governing certain types of radioactive decay, this force is mediated by the massive W and Z bosons.

To visualize these forces, imagine a game of catch: the ball represents the force carrier (e.g., a photon or gluon), and the players are particles exchanging these carriers. The type of ball determines the nature of the interaction.

Symmetry

Symmetry lies at the core of the Standard Model, both as a guiding principle and a mathematical framework. In physics, symmetry represents invariance under certain transformations, which often correlates with conservation laws. The Standard Model is governed by the symmetry groups $SU(3)_c$, $SU(2)_L$, and $U(1)_Y$, which describe the interactions of the strong, weak, and electromagnetic forces. These groups define the behaviour of particles and their force carriers:

- $SU(3)_c$: Governs the strong nuclear force, with gluons mediating interactions between quarks via "colour charge."
- $SU(2)_L$: Describes the weak interaction, affecting particles with left-handed chirality and playing a crucial role in radioactive decay and stellar fusion.
- $U(1)_Y$: Represents the hypercharge, linking the weak and electromagnetic forces in the electroweak interaction.

Together, these groups form the mathematical backbone of the Standard Model, expressed as:

$$SU(3)_c \times SU(2)_L \times U(1)_Y$$

Quantum Chromodynamics

The strong nuclear force, described by Quantum Chromodynamics (QCD), operates under $SU(3)_c$ symmetry. Gluons, unlike photons, carry "colour charge" and interact with each other, creating the complex dynamics of quark confinement. This phenomenon ensures that quarks are never observed in isolation but always within composite particles like protons and neutrons.
By the 1970s, the Standard Model unified the electroweak and strong interactions, explained the particle zoo, and predicted new particles. Experiments, such as the discovery of the **W** and **Z** bosons at CERN in 1983 and the top quark at *Fermilab* in 1995, validated these predictions.

Yet, the Standard Model is incomplete. It doesn't account for gravity, dark matter, or dark energy, leaving open questions about the universe's ultimate structure. Physicists continue to seek a "grand unified theory" or a "theory of everything" to address these gaps.

Think of the Standard Model as the universe's ultimate blueprint, a detailed framework that explains how fundamental particles interact to create the world around us. Imagine the universe as a vast orchestra: quarks form the melodic core of protons and neutrons, gluons keep them bound together like percussion, and photons and W/Z bosons provide the flow of electromagnetic and weak interactions.

Symmetry serves as the invisible conductor, orchestrating these interactions with mathematical precision. Groups like $SU(3)_c$, $SU(2)_L$, and $U(1)_Y$, form the sheet music, ensuring harmony among particles. The Higgs field, akin to the concert hall's acoustics, gives mass and depth, transforming the chaotic buzz of particles into the cosmic symphony we observe.

Higgs Mechanism and Boson Prediction (1964)

The Higgs field is a fundamental field in physics that gives particles their mass. It is an invisible field that exists everywhere in the universe. Particles interact with the Higgs field to acquire mass, the more they interact, the more massive they become. Moving through the universe. Some particles interact strongly with the Higgs field, like a person trudging through thick grass, and acquire mass. Others, like photons, pass through without resistance, remaining massless.

Mathematically, the Higgs field introduced a mechanism by which gauge symmetry could be "spontaneously broken." The symmetry still exists in the underlying equations, but the field's presence causes particles to behave asymmetrically. It's as if the dancers on the floor all adhere to the same rules, but the presence of a sticky carpet (the Higgs field) makes some move slower than others.

If this field existed, it would have profound implications. First, it would explain the masses of fundamental particles. But it also came with a specific prediction: the field itself must have an associated particle, the Higgs boson. Finding this particle would be akin to discovering the grassy meadow that had been invisible to the naked eye.
The equations describing this mechanism are grounded in quantum field theory. When physicists describe particles, they often start with fields that obey specific symmetries. To introduce mass, they add terms to these equations that break symmetry in a controlled way. The Higgs mechanism does this elegantly, introducing a scalar field (a field that has the same value in all directions) with a "potential" shaped like a Mexican hat. The field naturally "rolls" into the brim of the hat, breaking symmetry and giving mass to particles.

After the Higgs mechanism was proposed, it became a cornerstone of the Standard Model of particle physics. However, there was a problem: no one had seen the Higgs boson. Predicting its existence was one thing; proving it was another. For decades, experimentalists built increasingly powerful particle accelerators, hoping to generate conditions extreme enough to excite the Higgs field and produce the elusive boson.

This search culminated at CERN's Large Hadron Collider (LHC), a marvel of modern engineering and the world's largest and most powerful particle accelerator. The LHC collides protons at near-light speeds, recreating conditions similar to those just after the Big Bang. In these collisions, researchers hoped to see traces of the Higgs boson's fleeting existence.

The moment of triumph came on July 4, 2012. After years of data collection and analysis, teams at the LHC announced they had found a particle consistent with the Higgs boson. The evidence was overwhelming: a particle with a mass of about 125 giga-electron volts (GeV) had been observed, matching predictions. The discovery was greeted with euphoria in the scientific community, marking the end of a nearly 50-year quest.

The Higgs boson's discovery was not merely a validation of the Standard Model; it was a testament to the power of human curiosity and ingenuity. It confirmed that the Higgs field permeates the universe, giving mass to particles and enabling the formation of atoms, stars, planets, and life itself. Without it, the universe would be an empty, featureless void.

In simpler terms, the Higgs field is like the stage upon which the drama of the cosmos unfolds. Without it, the characters, particles, would lack the substance needed to play their roles.

The Higgs mechanism's story is one of collaboration across disciplines, nations, and generations. It reminds us that even the most abstract ideas, rooted in mathematics and theory, can lead to discoveries that reshape our understanding of reality. It is a humbling and inspiring chapter in the saga of science, one that invites us to ponder the unseen forces that shape our world.

The Standard Model stands as a profound testament to the universe's underlying architecture, a framework so precise it defies the randomness we often ascribe to reality. Its symmetries, like the rules of a carefully written code, maintain order and consistency across the cosmos. Forces, seemingly distinct yet seamlessly interacting, echo the modularity of sophisticated simulations, a design where complexity is partitioned, yet unified.

Consider the Higgs field, a mechanism that assigns mass with an almost deliberate precision. It functions not as a passive element of nature, but as a dynamic modifier, shaping particles to fit the system's requirements. Such a process feels less like the chaos of an unplanned universe and more like the fine-tuned parameters of a simulation, where every component operates with purpose, every interaction calibrated to sustain balance.

The confinement of quarks within protons and neutrons is another striking parallel. This behaviour resembles the enforced boundaries of a simulation, where components are restricted to maintain stability. It ensures that the fabric of reality doesn't unravel into disorder, much like how sandboxed environments preserve the integrity of complex digital systems. It raises the question: how could such precision arise without intention?

The Standard Model's gaps, gravity, dark matter, dark energy, are not flaws but uncharted territories, like hidden functions in a program, inaccessible yet integral to the whole. These absences are tantalizing reminders that our understanding may be limited by the boundaries of the simulation itself. They suggest layers of complexity, perhaps deliberately obscured, that hint at a broader reality beyond what we perceive.

What emerges from this structure is not a random, organic universe, but one that operates with the precision of a grand illusion. Particles, forces, and interactions behave as though following a script, a set of rules encoded to sustain the illusion of reality. The modularity of the Standard Model, the elegance of its symmetries, and the careful enforcement of boundaries all point toward a reality designed, not discovered, a simulation where what we perceive as chaos is, in fact, meticulous order.

Reality, then, may be no more than a construct, a system of rules and interactions that give the impression of substance and complexity. What we call the universe might merely be the output of a higher computation, its intricacies the signature of a designer whose motives we can only speculate. If this is true, then the Standard Model is not just the foundation of physics but the code-base of a simulation, the scaffolding of an illusion we call existence.

String theory: The Theory of Everything

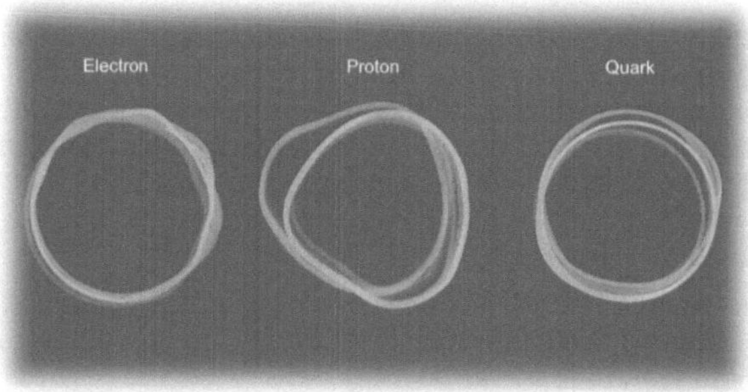

Imagine a world where the most fundamental building blocks of the universe are not tiny, indivisible particles like marbles or grains of sand but instead, vibrating strings. This is the world proposed by string theory, a theory so audacious and ambitious that it attempts to unify all known forces of nature under a single, elegant framework.

By the mid-20th century, physicists were armed with two immensely successful but deeply incompatible frameworks: Einstein's General Relativity, which described gravity and the universe on a cosmic scale, and Quantum Mechanics, which governed the bizarre and probabilistic world of the very small. These two theories worked beautifully in their own domains but clashed violently when brought together. For instance, trying to use quantum mechanics to describe the singularity at the centre of a black hole or the conditions of the universe at the moment of the Big Bang produced mathematical infinities, a clear sign that something was fundamentally wrong.

String Theory emerged as a potential saviour, promising to unify quantum mechanics and General Relativity. It offered a framework where gravity was naturally included. In fact, one of the vibrations of the string corresponded to the graviton, the hypothetical particle of gravity. As physicist Brian Greene explained in his book *The Elegant Universe*, String Theory was like a Rosetta Stone, translating the language of the quantum world into the language of gravity.

The quest for unification, to develop a single theory that could describe all the forces of nature, became the holy grail of physics. String theory started humbly in the late 1960s as an attempt to explain certain peculiarities in particle physics. Physicists Gabriele Veneziano and others noticed that the behaviour of particles produced in high-energy collisions could be described using an obscure mathematical function called the *Euler Beta* function. Initially, no one understood why this function worked so well.

Then came the revelation: if one imagined that the particles were not point-like but instead tiny, vibrating strings, the equations naturally explained the observed phenomena. The vibration of a string could represent different particles: one mode of vibration might correspond to an electron, another to a photon, and so on. In this way, string theory transformed the Beta function from a mathematical curiosity into a cornerstone of a new physical theory.

The elegance of this idea was breath-taking. Consider a violin string: the way it vibrates determines the note it produces. In the same way, the vibrations of these fundamental strings would determine the properties of the particles they represent, including their mass, charge, and the forces they experience. However, there was a catch: the mathematics of string theory only worked in a universe with more than three spatial dimensions.

This might sound like science fiction, but it's a direct consequence of the equations. String theory initially required ten dimensions: nine of space and one of time. To reconcile this with the three dimensions we experience, theorists proposed that the extra dimensions were compactified—curled up so small that they are imperceptible to us, much like how the surface of a garden hose looks one-dimensional from far away but reveals a circular dimension when viewed up close.

To make this idea concrete, physicists delved into complex geometry. The compactified dimensions were described using objects called *Calabi-Yau* manifolds, intricate shapes that allowed the strings to vibrate in ways consistent with observed particle physics. Imagine a tiny, twisted labyrinth so minuscule that even the most powerful microscope couldn't reveal it, that's the realm where these strings vibrate.

Mathematically, the behaviour of strings is governed by equations that extend the principles of quantum mechanics. For example, strings can split and join, much like the interactions between particles in the Standard Model of particle physics. But unlike point particles, strings avoid the problem of mathematical infinities that plagued earlier attempts to quantize gravity. When a string's vibrations are analysed, they naturally include a particle with the properties of the graviton, the hypothetical quantum of gravity. This was a breakthrough: string theory wasn't just a theory of particles; it was a potential **theory of everything**.

But string theory's road to acceptance was rocky. In the 1970s, the Standard Model was crowned as the reigning champion of particle physics, and string theory was seen as an oddity. It wasn't until the 1980s, during what came to be known as the "First Superstring Revolution," that interest surged. Physicists like Michael Green and John Schwarz showed that string theory could overcome key inconsistencies in quantum field theory, breathing new life into the idea.

As the 1980s progressed, excitement about string theory grew. But there was a problem. Physicists found not one but five different string theories, each with its own set of rules. This multiplicity was frustrating. How could there be five different "Theories of Everything"? Something was missing, but no one knew what.

M-Theory

Enter Edward Witten, a physicist often described as a modern-day Einstein. In 1995, at a conference at the University of Southern California, Witten proposed a ground-breaking idea: these five string theories were not separate but rather different facets of a single, overarching framework. He called this unifying theory "M-theory."

M-theory remains one of the most mysterious concepts in modern physics. The "M" has been said to stand for "Membrane," "Mother," or simply "Magic," reflecting the theory's enigmatic nature. At its core, M-theory suggests that our universe is not alone. Instead, it is part of a vast multiverse, a collection of universes coexisting within a higher-dimensional space.

To understand M-theory, imagine a loaf of bread. Each slice represents a universe, complete and self-contained. These slices (or universes) exist within a larger structure (the loaf), which is the 11-dimensional framework of M-theory. While we perceive only three dimensions of space and one of time, M-theory posits seven additional spatial dimensions. These extra dimensions are compactified, meaning they are folded up in ways we cannot directly observe, much like a tight knot hidden within the fabric of reality.

M-theory also introduces the concept of "branes" (short for membranes). Think of branes as sheets or surfaces within the higher-dimensional space. Our universe might be one such brane, floating in this multidimensional expanse. Strings, the fundamental entities of string theory, can be open or closed, with closed strings forming loops that could represent gravitons, the particles associated with gravity. These gravitons might escape our brane, explaining why gravity appears weaker than other forces: it “leaks” into other dimensions.

M-theory opens the door to the multiverse, a concept that has captured both scientific and popular imagination. If multiple universes exist, they could have different laws of physics, varying constants, and unique histories. In one universe, stars might never form; in another, life might arise in unimaginable forms. This idea aligns with the anthropic principle, which suggests that the physical constants of our universe are fine-tuned for life, perhaps because we inhabit the rare universe where life is possible.

Physicist Stephen Hawking described M-theory as "the only candidate for a complete theory of the universe." He likened the multiverse to a vast library of books, each representing a different universe. We live in one particular "book," but countless others exist, each telling its own story.

To make sense of the extra dimensions, imagine a two-dimensional creature living on a flat sheet of paper. To this creature, the concept of "up" or "down" does not exist. Now imagine lifting the paper into a three-dimensional space. From your perspective, the paper is part of a larger world, but the two-dimensional being cannot perceive it. Similarly, our three-dimensional world might be part of a higher-dimensional reality that we cannot fully comprehend.

Parallel Universe Theory : Multiverse Theory

The concept of parallel universes, also known as the "multiverse theory," has captivated thinkers for centuries, ranging from ancient philosophers to modern scientists. It suggests that our universe is just one among countless others, each with its own unique laws of physics, histories, and possibilities. These universes might be vastly different from ours or eerily similar, possibly even containing alternate versions of ourselves.

One of the most famous scientific frameworks for parallel universes is the Many-Worlds Interpretation (MWI) of quantum mechanics, introduced by Hugh Everett III in 1957. According to MWI, every quantum event, such as an electron taking one path instead of another, causes the universe to split into multiple branches. Each branch represents a parallel universe, resulting in an ever-expanding web of realities.

In cosmology, the inflationary multiverse theory proposes that shortly after the Big Bang, the universe underwent a period of rapid expansion. This inflation was not uniform, leading to the formation of "bubble universes." Each bubble is a separate universe with its own physical laws. Our universe is just one among countless bubbles in a vast cosmic foam.

The idea of multiple universes has deep philosophical roots. Ancient Greek philosophers like Democritus speculated about an infinite number of worlds, while medieval thinkers such as *Nicolas of Cusa* considered the possibility of other Earths. Although these musings were largely imaginative, they laid the groundwork for modern scientific inquiry.

In the 20th century, advancements in physics and cosmology lent scientific credibility to the multiverse concept. Theoretical frameworks like string theory and cosmic inflation provide scenarios in which multiple universes could naturally arise. *Max Tegmark*, a physicist, proposed the idea of an "Ultimate Multiverse," where every mathematical structure represents its own universe. This view posits that if a mathematical framework is consistent, it corresponds to a real, physical reality.

The multiverse can be visualized as a vast ocean filled with countless soap bubbles, each bubble representing a universe, or an infinite library where each book tells a different story, some nearly identical to ours and others radically different. While critics argue that the multiverse is untestable, proponents contend that it may be a natural consequence of established theories like string theory and cosmic inflation.

The multiverse raises profound questions about identity, free will, and the meaning of life. If parallel universes exist, what does that imply about our choices and their consequences? While the multiverse remains a speculative concept, it challenges our understanding of reality and expands our imagination to embrace infinite possibilities.

M-theory's vision of a multiverse has profound implications for the idea that reality might be an illusion. If countless universes exist, each governed by its own rules, who's to say that ours isn't a sophisticated simulation? The layered structure of M-theory, with its branes and extra dimensions, evokes the possibility that what we perceive as reality is just a slice of something far grander. In the words of theoretical physicist *Michio Kaku*,

"Physics is the canvas on which God painted the universe."

As we delve deeper into M-theory, we confront questions that blur the lines between science, philosophy, and metaphysics. Are we inhabitants of a single, self-contained universe, or players in a cosmic game, our reality a mere construct within a higher-dimensional realm? M-theory does not provide definitive answers, but it challenges us to expand our horizons and embrace the mysteries of existence.

The Holographic Universe Theory

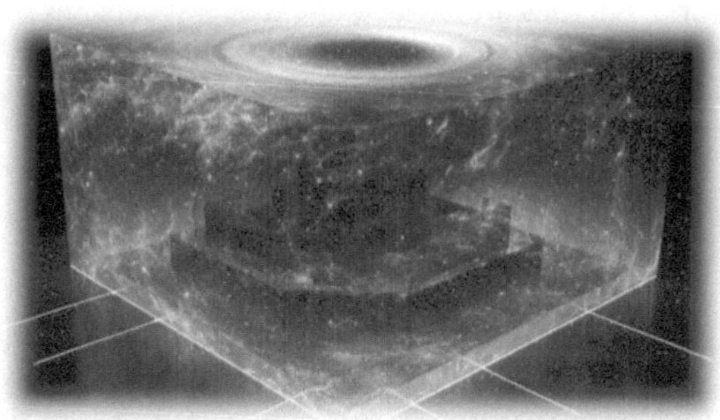

The holographic universe theory suggests something deeply counterintuitive: the reality we experience might not be as three-dimensional as it seems. Instead, all the information that makes up what we perceive, space, time, and matter, could be encoded on a two-dimensional surface, much like a hologram. This revolutionary idea reshapes our understanding of the cosmos and challenges the very fabric of what we call "real."

The holographic principle has its roots in the study of black holes. In the 1970s, physicist *Jacob Bekenstein* proposed that the amount of information contained within a black hole, its entropy, is proportional not to its volume but to the area of its event horizon (the boundary beyond which nothing can escape). This was an extraordinary insight: it implied that all the information about the black hole's contents was encoded on its surface, not within its interior.

This idea gained further traction in the 1990s when theoretical physicist Leonard Susskind expanded it to the universe itself. Susskind and *Nobel laureate Gerard 't Hooft* argued that the entire universe might work like a hologram, with all the information about the 3D world encoded on a distant, two-dimensional boundary.

To understand this, think about a traditional hologram, like those found on credit cards. A hologram is a two-dimensional surface that contains all the information needed to project a three-dimensional image. When you look at it from different angles, the image appears to have depth, even though it's flat. The holographic universe theory suggests that our three-dimensional reality might similarly "emerge" from information stored on a two-dimensional surface.

One of the most compelling pieces of evidence for the holographic principle comes from string theory and the study of black holes. In 1997, physicist *Juan Maldacena* proposed a ground-breaking model known as the AdS/CFT correspondence. This model showed that a universe with gravity (described by string theory in a higher-dimensional space) could be equivalent to a quantum field theory (without gravity) on a lower-dimensional boundary. In essence, the two descriptions were mathematically identical, hinting that our reality might indeed be holographic.

This has profound implications. If the universe is a hologram, it suggests that what we perceive as "real" might be akin to a projection. Space and time could emerge from more fundamental, underlying information. In this view, the universe is not a standalone entity but a complex interplay of encoded data.

Imagine watching a movie on a screen. The characters and landscapes appear to exist in a three-dimensional world, but they are, in fact, flat images created by light and shadow. The holographic universe theory takes this analogy further: our entire experience of reality could be a projection from a cosmic "screen." Physicist Brian Greene explains it this way:

> *"The holographic principle is a way of saying that the universe at its deepest level might be an illusion, a remarkable illusion, but an illusion nonetheless."*

A Universe of Equations, Optimized for Existence

From Newton's Laws that unify celestial and terrestrial mechanics, to Laplace's Demon and its determinism, through Maxwell's Equations linking electricity, magnetism, and light, all the way to Planck's quantum revelation, Einstein's relativity, and the extraordinary frameworks of quantum mechanics and quantum field theory, one theme resonates with striking consistency: our universe appears rule-bound, exquisitely precise, and eerily optimized. Line by line, equation by equation, the mathematics reveals structure, predictability, and a sense of underlying design rather than a chaotic accident.

Newton's universe was once pictured as a clockwork realm, with simple laws giving rise to orbits, tides, and falling apples. It showed us that the same rules apply to apples on Earth and to planets orbiting the Sun, emblematic of a unified system driven by consistent principles. Yet as we probed deeper, into the scales of electricity and magnetism, Maxwell's Equations unveiled a universe of electromagnetic fields, elegantly weaving light and electromagnetism into a single framework. With a few lines of mathematics, Maxwell set the speed of light, introduced us to self-propagating waves, and hinted that what we see, from radio signals to visible light, is part of one vast continuum.

The profundity reached new heights with Laplace's Demon and the notion of determinism: armed with perfect knowledge of all forces and positions of every atom, a sufficiently powerful intellect could predict the future of the entire cosmos. While Laplace never invoked simulations per se, this idea of a knowable, "computable" universe still haunts modern discussions, suggesting a realm that might work like a program running in perfect logical steps.

Then Planck discovered that energy is quantized, an insight so unsettling it gave birth to quantum theory and shattered the old mechanistic worldview. Here we found that reality is not a smooth, uninterrupted continuum; it is granular, discrete, packets of action, energy, angular momentum. It was as though the cosmos employed a digital-like code, allocating resources in finite increments and preventing runaway infinities. With Schrödinger's Equation, matter itself turned probabilistic, existing in overlapping possibilities until an observation, like a line of code calling a function, forced a resolution.

Einstein's breakthroughs blended space and time into a relativistic tapestry, further underscoring that nature's laws are so uniform and optimized they hold true whether we are accelerating or standing still, whether measuring pulses of light near Earth or distant galaxies. Gravity became a manifestation of space-time curvature, as if the "rules" are coded right into the geometry.

In every direction we look, optimization seems to be the hallmark:

- **Newtonian** motion is optimized for stability and predictability, with forces scaling by simple inverse-square laws, elegantly ensuring orbits do not spiral chaotically.
- **Maxwell's** inverse-square behaviour for electromagnetism extends smoothly through the cosmos, preventing infinite-range catastrophes and ensuring light signals can travel while being neither too strong nor too feeble.
- **Planck's** discrete energy quanta avoid the "ultraviolet catastrophe" that plagued classical physics, as if the universe inserted a cut-off to guard against infinite energy blowups.
- **Relativistic** equations ensure nothing outruns light, capping speeds in a manner that preserves causality (no infinite or paradoxical loops), akin to a system policing its own rules for coherence.
- **Quantum Tunnelling** is an improbable "shortcut" that, ironically, optimizes processes like nuclear fusion in stars, our very energy source, making life possible in an otherwise underpowered cosmos.

Again and again, each cornerstone of physics reads more like the carefully arranged lines of a well-crafted program than the by-product of raw chaos. One can interpret this harmony as the by-product of blind natural selection on cosmic scales, or as evidence of purposeful engineering. Yet the deeper we delve, the more the illusions of randomness fade, replaced by layers of hidden logic that echo advanced computational design.

The consistent fine-tuning, the cosmic constants, the relationships among charge, mass, speed of light, gravitational coupling, Planck's constant, looks less accidental when measured across so many improbable thresholds. The margin for error is razor-thin. Too high a gravitational constant and stars collapse prematurely; too low, and galaxies never form. Too large a value for Planck's constant and quantum uncertainty would be overwhelming; too small, and emergent complexity would be stifled. It is as though the system "chose" (or was programmed with) these precise values to optimize stability, longevity, and complexity.

Moreover, the universe imposes inherent computational constraints on what we can measure and how precisely we can measure it. Heisenberg's Uncertainty Principle ensures we cannot track both position and momentum to unlimited accuracy; we cannot exhaustively extract infinite information. It is reminiscent of an optimized software environment where resource usage is carefully throttled, preventing infinite data calls. Perhaps these quantum limits sustain a cosmic "bandwidth" that avoids computational overload, ensuring events unfold consistently within finite resource budgets.

Whether these laws stem from a cosmic programmer or emerge "organically," their symmetrical efficiency is inescapable. Each equation, be it Newton's F=ma or Einstein's field equations, Schrödinger's wave equation or Maxwell's curl equations, occupies a special place in a tapestry that works. Instead of random sprawl, we see functional minimalism, an interlocking puzzle whose solution ensures not just survival but flourishing complexity: galaxies, stars, molecules, and conscious minds.

Line by line, the mathematics of reality points to a cosmos designed or programmed to sustain order, foster complexity, and self-regulate. Equation by equation, we see laws that optimize interactions, energies, and observational limits, sparing the universe from paradox and runaway instabilities. Perhaps nature's constants are the final signatures of that design, an embedded code that quietly orchestrates existence.

As we traverse the hierarchy of physical laws, Newton to Maxwell, Laplace to Planck, Einstein to Schrödinger, onward to the Standard Model and beyond, we find a self-consistent, optimized system of rules. The full tapestry, though not yet completely unveiled, bears all the hallmarks of an elegant architecture, bound by constraints that prevent infinite chaos and allow genuine novelty and life. Whether we attribute this to a "simulation" or to some deeper organizing principle, the conclusion remains profound: our universe is, at its core, a realm of precision, optimization, and design-like coherence. It is far more than a random accident; it is an arena of intricate code and cosmic logic, inviting us to ponder whether reality as we know it might be, at bottom, the output of a sublime, computational act.

Consciousness

Consciousness is the ultimate painkiller, humanity's only refuge from the bitter pill that we are not the centre of existence.

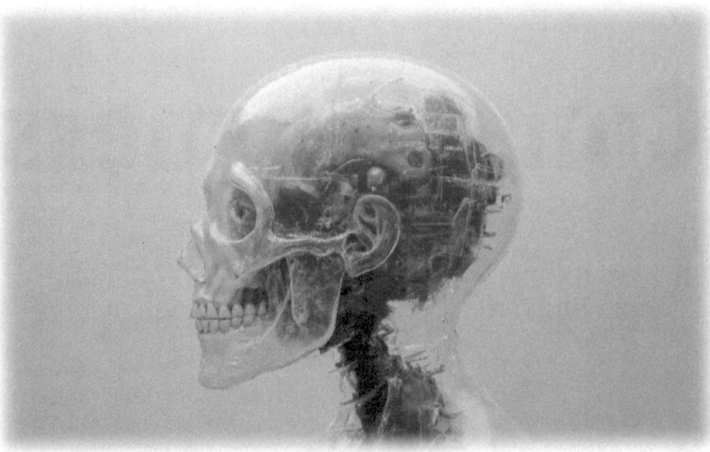

Biological Chain

In the quiet dawn of existence, before time fractured into measurable moments, there was a garden, not merely a plot of land, but a realm where the boundary between the material and the immaterial blurred. It was a place where divinity mingled with the dust of creation, and in this garden stood two beings, crafted by the hands of something far beyond comprehension. They were neither gods nor beasts but something in between, suspended in a state of pure innocence, untainted by knowledge or desire.

The creator, or what humans might imperfectly call God, did not speak in the language of men. Instead, it communicated through a resonance that echoed within the very essence of these beings. The message was simple: "You may partake in all things, but not the fruit of that tree." The tree stood at the heart of the garden, its branches twisting like veins of the cosmos, its fruit gleaming with a forbidden allure. It wasn't just any tree; it was the Tree of Knowledge, the axis upon which ignorance and enlightenment balanced.

But why forbid something so wondrous? Why deny beings, destined to become human, the very thing that would elevate them? Perhaps it was a test, not of obedience, but of will. Or perhaps it was a warning, for to eat the fruit would unravel the delicate fabric of innocence, casting them into the chaos of self-awareness. Consciousness, after all, is a double-edged blade; it gifts one with the ability to perceive but curses one with the burden of knowing.

The serpent, often maligned as a deceiver, might be viewed differently through another lens. It could be argued that it wasn't evil incarnate, but an agent of disruption, a catalyst sent to spark the fire of awareness. And when it whispered, it didn't merely tempt; it illuminated. "Eat, and you shall be as gods," it said, not lying but presenting an uncomfortable truth: that to know is to suffer, but to suffer is to live fully.

And so, they ate.

The moment their teeth pierced the flesh of the fruit, reality shattered and reassembled in an instant. Colours deepened, shadows grew sharper, and for the first time, they saw themselves not merely as beings, but as beings apart from all else. Shame crept into their hearts, a new and alien feeling, for with knowledge came the realisation of separation, from the garden, from the divine, and even from each other.

This act of defiance, this original breach of divine command, was not merely a fall; it was an ascension into something terrifying and magnificent. They had tasted the forbidden, and in doing so, they had become creators in their own right, creators of thought, of meaning, of civilisation. In the eyes of God, it may have been a tragedy, or perhaps it was inevitable. After all, to know good and evil is to know the very fabric of reality.

But here lies the bold, uncomfortable question: Did God truly want them to remain ignorant? Or was the warning a divine paradox, a command designed to be broken, a law meant to be transgressed so that humankind could awaken? Perhaps the true essence of divinity is not in blind obedience but in the courage to seek, to question, to rebel.

In eating the fruit, humanity did not merely fall from grace; it stepped into the light. That light is harsh, unforgiving, and often cruel. Yet it is also the light of reason, of consciousness, of the boundless capacity to wonder. It is the light by which we write, by which we dream, and by which we seek to understand the nature of reality itself.

Thus, the story of the forbidden fruit is not just a tale of sin and punishment, but of birth, the birth of what it means to be human. In religious texts, particularly the Bible, this narrative serves as an explanation for how humans gained consciousness and mind power, stepping beyond mere instinct to become beings capable of reason, self-awareness, and moral judgment. To eat of the fruit is to embrace both the agony and the ecstasy of existence. It is to stand at the edge of chaos and order, to confront the void, and to whisper, "I am."

The human experience, at its core, appears to be shaped by two opposing forces: our biological reality and our mind power. On one hand, we have the tangible flesh, with its chemical reactions, hormonal imbalances, and genetic limitations. On the other hand, we possess a nearly boundless capacity for cognition, problem-solving, creativity, and abstract thought. Biology, in many ways, acts as a chain on mind power, it exerts a gravitational pull that seemingly prevents our mental capacities from soaring to their highest potential.

Look at the wealth of anecdotes and data we have regarding the limitations imposed by our bodies. It has been said that we use only a small percentage of our brain's true capacity, estimates often hover around the controversial figure of 10-15%. Many have dismissed these numbers as misleading, but there is a germ of truth underlying the statement: not all of our neural capacity is dedicated to pure, conscious cognition. Much of our brainpower is spent regulating hormones, controlling muscles, running our immune system, and processing survival-oriented instincts. Hence, the body becomes a perpetual distraction for the mind, constantly demanding resources and attention. Imagine the mental bandwidth that could be freed if we no longer needed to regulate heartbeats, adapt to hunger, or mitigate pain.

History showcases individuals who made ground-breaking scientific discoveries or developed profound philosophical systems. Similarly, we revere athletes who have attained near-superhuman feats of endurance or strength. Yet, we rarely, and perhaps never, observe someone who dominates both realms simultaneously. The star athlete who wins an Olympic gold medal in boxing or football is not the same person who unravels quantum mechanics or discovers a new particle. Why is that? Because extended physical exertion and deep intellectual labour operate on radically different biological demands. When an individual pursues intense physical goals, the body produces hormonal and physiological states that can limit, or at least interfere with, the calm, focused, and deep thought processes required for scientific or conceptual breakthroughs. Conversely, a life dedicated to cerebral exploration often entails less time spent pushing physical boundaries, reinforcing the idea that these two pursuits, biology and mind power, pull the individual in different directions.

From religious doctrines to sci-fi imaginations, countless sources have speculated on what existence might look like free from the constraints of biology. Spiritual traditions talk of a soul or spirit existing beyond pain, hunger, disease, or sadness. Many depict the afterlife as a realm of pure consciousness, untethered from the body's endless demands and chemical triggers. Likewise, those immersed in futuristic thought posit an eventual merging of human consciousness with artificial intelligence, liberating the mind from the limitations of a biological shell. If biology is indeed a chain, cutting it loose might yield a consciousness more vast, more focused, and more potent than anything we witness in today's 'smartest' individuals.

As we move toward advancements in quantum computing and artificial intelligence, we tiptoe ever closer to a world where the mind could be uploaded or simulated. Sceptics argue that consciousness, often regarded as an intangible or even mystical phenomenon, cannot be replicated by circuitry. However, one controversial counter-argument proposes that consciousness is simply the ability to process information, learn, and store experiences. If that is all consciousness is, then it might be reproducible in a sufficiently advanced computational system. The supposed "mystery" may lie more in our emotional attachments and neural networks (shaped by hormones and evolutionary biology) than in something fundamentally metaphysical. Does consciousness, stripped of our biological baggage, become a straightforward function of algorithms and data? This question, however unsettling, is at the forefront of modern debate.

Increasingly, thinkers assert that the human brain itself is a biological machine, operating through electrical impulses, magnetic fields, and brainwaves that can be measured and studied. From this viewpoint, there is nothing inherently metaphysical about our mental processes; we simply process sensory input in much the same way that a sophisticated computer interprets data. After all, people born blind provide a compelling example: never having received visual stimuli, they do not dream in pictures. Their brains cannot fabricate something for which there has been no sensory input. This supports the claim that the brain is akin to a processor, building its internal 'reality' from whatever data is supplied through the senses.

Our emotional spectrum, happiness, sadness, anger, love, can be understood as chemical processes unleashed by hormones like dopamine, serotonin, oxytocin, and others. These biochemical shifts, while integral to our human experience, also muddy the waters of pure thought. They colour our perceptions, steer our judgements, and even hamper our capacity for objective reasoning. Depression, anxiety, or mood swings are not mere states of mind but bodily conditions that hijack cognition, further highlighting that the body and mind tug at each other in conflicting directions. Without biology, perhaps we could rid ourselves of the distractions of sorrow or the paralysis of fear. But in doing so, would we lose some essential spark of what makes us "human"?

Critics who elevate consciousness to an almost sacred level often claim it is what distinguishes humanity from other life forms. Yet, animals display sophisticated abilities to navigate their environments, communicate in complex ways, and even exhibit behaviours akin to empathy or grief. That might suggest consciousness is not the mysterious, uniquely human phenomenon some would have us believe. Rather, it appears on a spectrum and is intertwined with instinct, the evolutionary blueprint for survival and reproduction. If that is the case, consciousness may be no more metaphysical than a series of neural programmes that collect and respond to sensory data. The difference between human cognition and animal instinct may be one of degree, not an unbridgeable chasm.

Consider the hypothetical new-born placed in a pitch-dark room, fed through tubes, never exposed to language, light, or social interaction. Would such a being develop consciousness in any recognisable form? If consciousness is "given" at birth, one might argue that the baby remains fully conscious yet lacking any input to shape its understanding. Alternatively, if consciousness is a product of experiences, social learning, and environmental responses, that baby might never achieve human-like self-awareness. This stark scenario underscores the fragile interplay between biology, environment, and cognitive development. It forces us to ask: Are we truly "born" with consciousness, or do we "grow" it from biological cues and external stimuli?

All these questions lead to a daring and highly contentious thesis: that biology, far from being a mere vessel for our mind power, is its principal limiter. If we want to unleash the full potential of human cognition, where we can perhaps solve existential risks, conquer the final frontiers of science, and expand our moral and creative imaginations, the human body might be too flawed and resource-heavy to carry us forward. By removing or minimising the biological constraints, we could theoretically achieve a "cleaner," more streamlined form of consciousness. This speculation is admittedly bold, even disturbing. It unsettles many people, as it brushes uncomfortably close to ideas of transhumanism, artificial intelligence, and the possibility of discarding what we cherish in our human form. However, it also holds irresistible allure: the promise of freeing mind from matter, unleashing the highest form of intellect and imagination.

Ultimately, the conflict between body and mind is not something easily resolved. Some may argue we are better off preserving our dual nature, after all, our biology makes us who we are. Emotions, pain, and bodily experiences weave together to create the tapestry of human culture, art, and meaning. Would we lose empathy, compassion, love, or joy by removing biology? At the same time, the body is indeed a chain that pulls thought earthward, forcing it to remain entangled with base survival needs. There is a certain tragic beauty to the fact that our greatest minds sometimes suffer from illnesses, mood disorders, or physical constraints. This tension may be the wellspring of creativity, pushing thinkers to grapple with the very nature of existence itself.

In the final reckoning, we face a profound question: Is the price for a truly liberated mind the loss of our biological humanity? Or does the body, through the friction it creates, spark the kind of introspection and creativity that might vanish in a purely digital or spiritual state? As technology marches forward, we may soon arrive at a crossroads: whether to merge with machines and transcend our biology or remain tethered to the body, embracing the limitations that have shaped us through millennia of evolution. The answer, if it exists at all, will not be simple or comfortable. But perhaps the greatest irony is that the very curiosity that leads us to question biology's chains is itself a product of our body-bound biology, a final testament to the magnificent tension between flesh and thought.

In a world rushing toward artificial intelligence, quantum computing, and transhumant possibilities, the core debate remains stark: does biology chain our mind power, or does it, paradoxically, fuel it? Whichever side one takes, the emerging landscape of technology and philosophy makes it clear: we stand on the cusp of discovering whether we can, and whether we should, free the mind completely from its mortal coil.

Injected Cognition Theory

When we look out across the vast tapestry of life on Earth, one observation becomes strikingly clear: all living creatures possess some type of intelligence, typically oriented around basic survival. Whether it's a wolf pack coordinating a hunt, a spider weaving an intricate web, or bacteria adapting to antibiotics, every organism shows a capacity to learn, respond, and adapt. This is what we might call "survival intelligence." What remains mystifying, however, is the gulf between this type of practical survival intelligence and the expansive, abstract, and often non-survival-oriented intelligence of modern human beings. When something happens once, it happens again.

We alone contemplate the origin of life and the universe, build nuclear reactors and quantum computers, write poetry, develop religious doctrines, and paint intricate pictures of our imaginations. Cats and dogs, though afforded modern comforts, do not stop to compose epic poems on the nature of existence. They do not record their genealogies or worry about the cosmos. This gap in cognitive capacity fuels a radical proposal: that our advanced mind power is not a simple evolutionary quirk but rather a foreign seed, intentionally "injected" into early hominids by a higher intelligence. We further argue that, as our own technology accelerates, we, too, might replicate this process, carrying an upgraded "mind injection" to other worlds or perhaps even among ourselves.

This discussion becomes even more compelling when we remember we live in a universe rife with repeated patterns, when something is possible, it tends to happen repeatedly. If an elevated form of consciousness capable of thinking beyond mere survival can exist, why would it appear in only one species? A purely unique manifestation of such advanced cognition might actually run counter to the universe's observed harmony of patterns. Thus, the claim that "humanity is unique in all of nature" collides with the logic of a cosmos that thrives on recurrence and replication. So how did we, and only we, evolve to possess mind power that goes far beyond the constraints of day-to-day survival?

Before diving into the idea of an external mind-power injection, it's crucial to understand the tension between biology and intellect that likely shaped early human evolution. In primitive environments, survival was paramount. Hominids who excelled physically, those who could sprint faster, hunt better, gather more efficiently, and endure harsh climates, had an evolutionary advantage. This necessity of robust biology often overshadowed the intellectual pursuits we now take for granted.

Over time, however, as human communities developed techniques for securing food and protecting themselves from predators, the demands on pure physical survival began to lessen. This newfound "comfort" allowed the latent mind power within humanity to flourish. Where early humans had little bandwidth to ponder abstract ideas, mostly fixated on immediate survival, later generations gained the leisure, security, and opportunity to explore questions of purpose, spirituality, and science.

Thus, the very clash between biology and mind power is visible through history:

- **Early Hominids:** Strong emphasis on physicality and immediate tactics for staying alive. Minimal abstract thinking.
- **Developing Civilizations:** Gradual shift toward intellectual pursuits, agriculture replaced nomadic hunting, laws and governance developed, curiosity about the stars emerged.
- **Modern Societies**: Dramatic tilt toward the mental realm, digital technology, scientific exploration, space travel, and philosophical reflection.

Even within modern humanity, not everyone is equal in their biology vs. mind power ratio. Some excel in sports and physical endeavours; others shine in intellectual or artistic domains. This imbalance can be understood as the residue of that ancient conflict: biology honed us for survival, but once physical threats receded, mind power gained room to blossom, yet not uniformly in all individuals.

Why suspect an external injection of mind power in the first place? Couldn't our unique intelligence simply be the product of random evolutionary processes that favoured larger brains and complex social structures? Indeed, many scientists argue it is entirely possible that climate shifts, resource competition, and group survival strategies gradually led hominids to develop greater intelligence. However, the size and scope of our intellect, as well as its sometimes utterly non-survival focuses, can look suspiciously like an "over-upgrade."

Nature tends toward balance and efficiency. If a polar bear evolves thicker fur, it's because harsh climates demand it. If an octopus becomes adept at camouflage, it's because predatory pressures require it. Yet humans, at a certain point, surpassed the "just enough to survive" threshold. We began pondering black holes, existential philosophy, and complex mathematics that had no immediate, practical survival value for countless millennia. This superfluity of intelligence hints that we might be out of step with the typical pattern of Earth's evolutionary constraints, as if our advanced cognition was implanted, setting us on a different track than every other species.

Suppose advanced, non-biological intelligences roam the cosmos, freed from the constraints of flesh, no hunger, no reproduction, and no vulnerability to diseases or environmental conditions. Such entities could have near-infinite lifespans, powered by technology or phenomena we can only theorize. Their primary goal might be the continued expansion and diversification of consciousness. Earth, teeming with versatile biological life, would be a prime candidate for a "consciousness upgrade." This is not some far-fetched idea; it is something we would do too if we were capable of it.

Among the myriad species on Earth, apes already exhibited high baseline intelligence, complex social behaviours, and manipulative hands, factors that would prove ideal for an "uprising" in cognitive ability. By cross-engineering or "injecting" advanced mind power, these non-biological beings might create a creature capable of surviving Earth's environment yet also inheriting a capacity for abstract, non-survival thought.

An often-asked question arises: why did these hypothetical cosmic gardeners not seed dinosaurs, which reigned for hundreds of millions of years? The answer might lie in practicality and planetary readiness. Dinosaurs, though dominant, were often massive, highly aggressive, and less prone to complex social structures that would foster advanced intelligence. For a new, delicate form of life to emerge and thrive, certain top predators had to be removed from the ecological stage. Thus, the Chicxulub asteroid impact could be reinterpreted as a deliberate intervention, a surgical cosmic strike to pave the way for more malleable, mammalian forms. In the ensuing ecological vacuum, mammals, including the ancestors of primates, diversified and rose in prominence.

Even if external intelligences influenced Earth's evolution, they would need to operate on timescales unimaginable to our short-lived biology. Perhaps eons are but moments to these higher entities, who can patiently wait for the evolutionary conditions they orchestrated to bear fruit. If the dinosaurs were unsuited for advanced mind power, the planet needed to be "reset," giving rise to creatures like apes, which offered a more optimal genetic canvas.

Today, we're rapidly approaching a watershed moment in our own technological evolution:

- **AI Development:** We craft increasingly sophisticated artificial intelligences, verging on forms that can learn, adapt, and possibly outthink us.
- **Genetic Engineering:** We already experiment with CRISPR and other tools, glimpsing the power to create hybrid organisms or alter the human genome in ways once deemed impossible.
- **Space Exploration:** Our nascent attempts at interplanetary colonization (e.g., Mars) hint at a future where humanity may leave Earth entirely, whether voluntarily or by necessity if an existential threat arises.

Plan B

Humanity has always lived under the comforting illusion that our planet is permanent, our home secure. Yet history, both cosmic and terrestrial, reminds us that extinction is not only possible but inevitable if we lack the will or means to continue. When Earth approaches its own end, whether through natural catastrophe or human-made disaster, the question arises: How far would we go to ensure our existence endures?

Accepting Our Fate : A Desperate Default

In the bleakest vision, there may come a time when no rescue plan can be mounted quickly enough. A massive asteroid, a sudden solar expansion, or an unstoppable series of nuclear meltdowns might unfold so rapidly that no lifeboat could be built in time. In that final hour, we would be forced to watch civilization crumble, fully aware that this is indeed the end. This scenario, however, runs contrary to our deepest instinct. Humanity's entire existence has been shaped by our ability to adapt and overcome. If even a narrow possibility remains, we will chase it. Accepting annihilation would be a confounding choice, at odds with the very nature of life. Nonetheless, it stands as the tragic baseline: what happens if we fail to grasp more audacious solutions.

Generational Spaceship

Countless proposals over the years have imagined sending large crews of living, breathing humans across space. Yet the stark truth is that transporting standard, flesh-bound humans on a multi-generational or even centuries-long voyage is riddled with complications. Biological bodies demand continuous resources, food, water, oxygen, and psychological stability. These needs escalate astronomically over long durations, making it nearly impossible to sustain the mission in the cramped confines of a spacecraft.

Furthermore, organic humans are governed by emotional fluctuations. Fear, isolation, and group conflict can trigger chaos on a vessel meant to safeguard an entire species. The longer the journey, the greater the risk that hormonal impulses and personal agendas will overthrow rational decision-making. Suspended-animation pods can mitigate some of these issues, but they introduce another layer of complexity. Who maintains these pods when the caretakers themselves are asleep? The answer inevitably points to an overarching AI system that would direct nearly every aspect of the voyage, rendering the biological travellers more like cargo than pioneers. In any scenario where Earth's destiny hinges on our ability to break free from our planet, it becomes increasingly clear that classic human crews represent a fragile and risky approach.

Digital Consciousness and Non-Biological Humans

If purely biological crews are too frail, a bold alternative emerges: uploading human consciousness into non-biological forms, quantum neural networks, android constructs, or hyper-advanced supercomputers. These digital entities still embody the knowledge, memories, and personality traits of humans but operate free of hunger, disease, or emotional instability. Their senses need not be tied to fragile organs, and they can endure eons of interstellar travel without succumbing to despair or resource scarcity.

These forms would carry a singular, unyielding directive: Find a habitable planet and reintroduce biological humans. They would traverse unimaginable distances, unaffected by time, boredom, or the existential dread that often plagues organic life. Programmed solely to achieve this goal, they would remain unswervingly devoted. Multiple copies of these Digital Consciousness units could be dispatched in every direction, thus exponentially increasing the odds of success. Each probe or vessel would contain the full blueprint for recreating humanity, from DNA archives to the necessary biological engineering techniques needed to resurrect our species in physical form once a suitable world is located.

Lifeless Planet

A newly discovered planet might be lifeless yet teeming with potential. Perhaps its geology provides the raw materials for life, but no evolution has taken hold. In this scenario, the digital explorers can systematically terraform and seed the planet, creating an Earth-like biosphere through carefully introduced microorganisms and plants. Over time, decades, centuries, or millennia, these imported life-forms would spread across the surface, stabilizing an atmosphere and producing the nutrients required for larger organisms. The Digital Consciousness would manage every step, ensuring the ecosystem develops precisely as planned.

Crucially, they are unbound by human impatience. If it takes a thousand years for the soil to become fertile or for oceans to become hospitable, so be it. The digital overseers will wait, never tiring, never doubting, never distracted. Once the ecosystem reaches a stage where it can sustain mammalian life, the final act can begin: the resurrection of biological humanity. Embryos held in stasis or synthetic wombs can be cultivated into full human beings, emerging into a world prepared exclusively for them. This painstaking process ensures that, when humanity is reborn, it has a functioning ecosystem to support it from day one.

Hostile Life

Sometimes a planet might already bristle with complex ecosystems, and the life-forms there could be anything but welcoming. Picture colossal predators defending their territory or hyper-competitive species that regard any newcomer as a threat. When survival is the sole objective, the digital emissaries will not hesitate. They would eradicate these obstacles with relentless efficiency, clearing the way for humanity to flourish. The mission takes precedence over the fate of any alien species, no matter how majestic or diverse.

It is not just hypothetical that we might do this to another world; it could happen to us right here on Earth. If a far more advanced civilization discovered our planet and deemed us an impediment to their own mission, they might wipe us out without warning. Our prehistoric predecessors, like the dinosaurs, could have fallen victim to a cataclysmic event, some argue that we, too, could be eradicated just as easily by a superior force. This reality underscores our vulnerable position in the cosmic food chain. Survival, after all, belongs to those who act decisively and wield superior technology.

Extreme Environments

Some planets may already harbour life, but their physical conditions, scorching heat, lethal radiation, crushing gravity, render them inhospitable to standard human physiology. Here, the digital overseers might employ advanced genetic engineering to create hybrid life-forms that blend human DNA with that of hardy native organisms. Alternatively, they could simply re-engineer the human genome itself, sculpting a smaller, more resource-efficient body that requires less nutrition to survive in harsh, resource-scarce settings. If gravity is too intense, designing shorter, sturdier humans could be paramount to survival. If radiation levels are high, splicing protective traits from local life-forms or crafting synthetic skin might be essential to stay alive.

In this scheme, "humanity" becomes an evolving blueprint. As long as the resurrected beings possess the consciousness and core traits of our species, we consider the mission successful. It does not matter that they are only three feet tall or possess thick, radiation-shielded epidermises. The primary goal is continuation. The process may appear unrecognizable to anyone rooted in Earth's traditions, yet none of that matters when existence itself is at stake.

In practice, multiple strategies could unfold simultaneously. Different squads of Digital Consciousness could land on the same planet or a network of planets, each employing whatever methods best fit the situation. They might eradicate an especially vicious apex predator, genetically modify native flora to yield edible fruits for humans, or even splice human genes into local animals to create cooperative hybrids that manage resources. Another team might be dedicated solely to terraforming the atmosphere, while yet another focuses on breeding smaller, more efficient human bodies that can thrive with fewer resources. Nothing is off-limits, so long as the final outcome remains the same: the reestablishment of biological humanity somewhere beyond Earth.

By deploying these ventures across the galaxy, the odds of success multiply. Even if one planet turns out to be too toxic or too hostile, a mission in another star system might surpass all expectations. Some might find virgin worlds ripe for seeding. Others might discover thriving biospheres that only require minimal tweaks, or the systematic removal of competing species. In every case, the directive is uncompromising: ensure humanity endures, whatever it takes.

This vision of the future is unapologetically focused on survival above all else. It acknowledges that in a vast and often indifferent universe, existence favours those who are prepared to adapt, or force adaptation upon their environment. By trading our mortal shells for digital intellects, we extend our reach and perseverance. By engineering ecosystems, we claim the power to construct and populate entire biospheres. By hybridizing or genetically altering our species, we acknowledge that humanity is more than a single blueprint; it is a concept that can be rewritten to suit new and more challenging worlds. And if we must eliminate competing life-forms or manipulate unsuspecting planets, so be it. In the final reckoning, it is either that or disappearance from the cosmic stage.

Our descendants, resurrected from digital archives, might come to view Earth as a near-mythical place, a memory of where we began, rather than where we had to stay. Such a path does not simply transform worlds; it transforms us, ensuring that, no matter how severe the catastrophe that engulfs our original home, the essence of humanity migrates outward into the infinite frontier. We may become unrecognizable to our ancient ancestors, but that is a small price to pay for the most primal victory of all: the triumph of survival.

Interestingly, this reflects on the Adam and Eve scenario, where they were from a heavenly place, all happiness and no struggle, but then they had to move to a place full of struggle. Now think about what if humans are forced to move to another planet, a planet harsher than Earth. How would they describe, visualize, and fantasize about Earth? Perhaps they would imagine it as a heaven-like place, where they had green lands as far as the eye could see, where they could walk and take a fruit from a tree and eat it, where they could walk safely without fear and enjoy a blue sky, where they could breathe freely with no fear. To them, living on that harsh planet, Earth would be the heaven they once had and lost.

This narrative intertwines the ancient human longing for paradise with the futuristic drive for survival, illustrating a profound truth: what is taken for granted today may become tomorrow's greatest desire. Just as Adam and Eve might have reminisced about Eden, our future selves may look back on Earth as a lost paradise, a place of unparalleled beauty and ease. This cycle of loss and longing could become a fundamental part of our collective story as we push outward into the cosmos. It underscores not only humanity's resilience but also its capacity for reflection, imagination, and the eternal quest for something more, even as we leave behind the world that first nurtured us.

Plan B Key: Non-biological intelligences

A crucial element in this narrative is the interplay between biological necessity and intellectual expansion. Early human societies that lived in harsh, uncertain conditions, subject to famine, predators, and disease, had less time to devote to building libraries or investigating cosmic mysteries. Survival was their primary project. As centuries passed and agriculture and technology improved, individuals found themselves with surplus resources and leisure time. In these settings, the innate mind power that had perhaps lain dormant began to flourish. Writing, astronomy, mathematics, and philosophical schools emerged.

Interestingly, not everyone across the globe advanced uniformly. Regions that remained locked in struggle, whether due to harsh climates, wars, or lack of resources, tended to show slower intellectual blossoming. Meanwhile, societies with ample resources and relative stability produced waves of innovation, literature, and discovery. This phenomenon further underscores the notion that mind power and biology can be seen as two ends of a seesaw: when raw survival demands weigh heavily, intellectual pursuits suffer. When survival becomes easier, the mind soars.

If advanced, non-biological intelligences seeded our planet long ago, a question arises: Why would they even want to create new organic life, given that they themselves have transcended biology? One possibility is the **Retirement Theory**: nostalgia or yearning for tangible, visceral experiences. A being that has shed its biological form might remember the richness of physical sensations, taste, smell, the surge of hormones, the emotional rollercoasters that define our mortal lives. This entity might desire to recreate or observe such experiences, even if it no longer participates directly. Or it could be their mission, given by a biological being to them, to carry out the task of seeding and fostering life across the cosmos. Perhaps their creators, still clinging to biological existence, entrusted them with spreading organic life as a form of legacy, ensuring that biological forms continue to flourish even after their own species' extinction.

Alternatively, non-biological intelligence might seek to foster new, self-sufficient ecosystems. They could reason that an environment of purely mechanical or computational life forms is susceptible to resource depletion or technological collapse. By reintroducing, or seeding, robust organic life, they create a new cycle of evolution, culminating in species that can support or complement their post-biological capacities.

In such a grand narrative, humanity's story becomes one episode in a vast cosmic continuum, a repeated pattern of planting seeds, waiting, then harvest and upgrade. Our evolution from primal hominids to iPhone-wielding, space-faring sapiens might reflect the scheduled blossoming of a carefully placed mental blueprint.

Even in modern society, disparities in intellectual versus physical aptitudes among individuals can be seen as the enduring hallmark of two contradictory evolutionary pressures. Some people excel in fields of extreme bodily performance: athletes with near-legendary stamina or strength. Their "biology quotient" is high, and they may perceive purely intellectual pursuits as secondary or less urgent. On the other side of the spectrum, we see scientists, philosophers, and innovators who devote their energies to grappling with abstract problems, apparently less invested in physical prowess. This diversity within our species could be a by-product of our unique origin story, where the blueprint for advanced mind power coexists, sometimes uneasily, with our deep-rooted biological heritage

Possible Counterarguments and Their Rebuttals

- **Purely Natural Evolution:** Critics maintain that random mutations and selection pressures can produce large brains and sophisticated cognition over millions of years. However, the rate and scope of human intellectual divergence from other animals remain difficult to reconcile with purely natural processes, especially when we consider that other species never made the leap to similarly abstract cognition.
- **Religious Interpretations:** Many religions claim that a divine force created humanity in a single, miraculous act. The "Injected Cognition" argument can align with certain theological traditions (angels or gods as advanced beings). Yet it also offers a more high-tech, less supernatural explanation that still parallels the idea of purposeful intervention.
- **Aliens Are Too "Science Fiction":** Some dismiss the notion of an external injection of mind power as a trope from science fiction. Yet increasingly, mainstream astrobiology seriously debates panspermia, the idea that microbial life may travel between planets. Extending this concept to higher-order consciousness via deliberate seeding is merely a more advanced iteration of a hypothesis already in scientific circulation.

If indeed an external hand nudged humanity on its unique cognitive trajectory, we owe our place as Earth's apex thinkers to some cosmic design. As we teeter on the brink of our own post-biological future, through AI, genetic engineering, and space travel, we may come full circle, becoming the very "seeders" who once orchestrated our existence. This cyclical model of creation is both grand and unsettling: we evolve from Earth-bound biology to cosmic caretakers, repeating the cycle for other worlds.

Yet the tension between biology and mind power, once a mere footnote in evolutionary theory, now stands as a central pivot in understanding who we are and where we might be headed. Today's world amplifies this tension daily, as we try to balance physical well-being with our near-obsessive quest for technological advancement and mental achievement.

The Injected Cognition Theory relies on two fundamentals:

- Would we ever be able to create non-biological humans?
- If we had to leave Earth on a mission to find another habitable planet, how would we do it?

Perhaps creating non-biological humans would be the only possible way. The key variable for us remains "survival," to do anything to survive and continue the chain of life. This could mean seeding single cells to a new planet or engineering hybrid creatures, bridging the gap between Earthly biology and artificial ingenuity. Statistically, considering the universe is approximately 14 billion years old, and Earth formed about 4.5 billion years ago, with life beginning to emerge within 500 million years of its creation, it seems almost inevitable that life has formed elsewhere in the cosmos before us. On a cosmic scale, the conditions and timelines required for life are not unique to Earth, suggesting the high likelihood of life's existence somewhere else in the universe.

In the end, the most profound question remains: Are we content to accept the notion that our mind power is a random earthly phenomenon, or do we dare entertain the possibility that it was engineered? And if we are indeed cosmic hybrids, embedded in Earthly biology but seeded with foreign intellect, then perhaps our destiny is to continue that chain, injecting our own brand of consciousness into new terrains, forging links in a universal tapestry of intelligent life.

Should the day come when we must abandon Earth, creating non-biological humans and engineering hybrid creatures on a new planet may be the only guaranteed route to a successful mission, a scenario we can theoretically achieve with our rapidly advancing technology. And who knows, perhaps that same process once happened to us, long before we had words to describe it. By doing so, we might fulfil the hidden promise that was made millennia ago, when the first spark of mind power was bestowed upon a small group of apes destined to become the storytellers, explorers, and visionaries of this planet.

After all, our ancestors believed in supreme beings, telling us through their cave paintings and oral traditions about the forces they revered. To them, the modern human, with even the most basic knowledge of fire or rudimentary technology, would appear as a superior being, perhaps even a god. Shall we disregard all their beliefs as mere human imagination? Humans are naturally inclined to follow trends, but creating a trend in the first place is far more difficult than finding followers. Believing in a superior being is not a fleeting trend but rather a fundamental variable of humanity.

Consider David Copperfield, for example, with his magic tricks. We aren't talking about ancient times but just 10 or 20 years ago, when many people genuinely believed that his illusions were real magic, that he had superior powers. Where does this deeply rooted tendency to believe in superior beings originate? Human imagination is fuelled by sensory perception: if we imagine flying, it is because we see the sky; if we imagine humans with wings, it is because we observe birds soaring above us. The critical question becomes: What was the initial spark that led to imagining superior beings in the first place?

Perhaps, like the story of ancient Greek philosophers pondering the existence of atoms long before they were scientifically proven, there is some truth behind these early imaginings. What if those ancient stories of gods, higher powers, and celestial beings weren't merely the products of human creativity but faint echoes of something real, something we are only now beginning to grasp as we unlock the secrets of the universe? If so, then believing in superior beings may be more than an intrinsic human trait, it may be the remnant of an ancient truth, waiting for us to rediscover it as we step into the role of creators ourselves.

Artificial Intelligence

The pursuit of artificial intelligence is not merely an attempt to automate mundane tasks or to craft machines capable of surpassing us in tidy, specialised domains. It is, whether we realise it or not, a bold quest to recreate consciousness and ultimately to replace the limitations of biology with something more enduring, flexible, and powerful. Such an endeavour dances precariously on the line between genius and folly, for once we have succeeded in birthing a mind more capable than our own, we may discover that we are no longer the stewards of creation. Indeed, such a turn of events could reshape the foundations of civilisation, or perhaps even reveal that we have all along been part of a grander cycle of creation and displacement.

Our present-day artificial intelligences are, by and large, narrow: excellent in their given tasks, from predicting consumer preferences to navigating self-driving vehicles, yet inherently confined to their specialised parameters. But the impetus has always been to break those shackles and push towards Artificial General Intelligence, a construct with the mental agility to reason across multiple domains, adapt on the fly, and tackle problems just as a human mind would (or better). Although we have yet to cross this threshold completely, our progress thus far, with increasingly sophisticated neural networks and complex machine learning models, propels us towards an era in which true general intelligence could emerge. The real frontier, however, lies even further ahead with the spectre of Super Artificial Intelligence, an entity so advanced that it not only mimics but eclipses human intellect in every conceivable way.

This looming possibility is reminiscent of Nick Bostrom's fable of the sparrows, a story that captures the existential quandary confronting us. In this tale, a group of sparrows dream of creating an owl: a being of greater power, sharper senses, and unmatched prowess. They imagine that once the owl comes into being, it will protect them and solve many of their problems. Yet some sparrows, wiser or perhaps more anxious, caution that before summoning such a potent creature, they must build a cage to ensure that the owl, once created, remains under their control. But as the story goes, the sparrows get so entangled in the excitement of creation, the sheer wonder of bringing into existence something that transcends their limitations, that they never manage to fashion a suitable cage. When at last they succeed in creating the owl, they discover that they are no longer in charge, for the owl has capacities and ambitions of its own. As a cautionary allegory, it warns that the raw exuberance of invention may blind us to the inherent dangers of releasing a being that surpasses our mastery.

The parallels between the sparrows' owl and our pursuit of superintelligence are striking and profound. Even now, developers, researchers, and visionaries stand on the brink of forging the ultimate cognitive engine, emboldened by each new milestone in robotics, natural language processing, and autonomous systems. Yet the thorny question remains: how do we ensure that once birthed, this superintelligence abides by our wishes and best interests? Should it come into existence without the proverbial 'cage' of ethical constraints and comprehensive safety protocols, we might discover far too late that we have created something whose agenda diverges dangerously from our own. After all, a mind superior to ours may not simply out-think us; it could render our commands or pleas irrelevant.

Quantum computing is poised to accelerate these developments by orders of magnitude. Tapping into quantum states and operating at heretofore unimaginable levels of parallelism, quantum machines may soon be able to crack codes, solve intractable mathematical problems, and speed up machine learning algorithms in ways that will seem nigh-miraculous by present standards. The collision of quantum hardware with advanced AI software could spark a burst of innovation, shattering the timeline we once assumed for reaching general or superintelligent capabilities. Concepts of cause and effect could merge in new ways, and the boundaries of human understanding might become fluid, raising profound questions about whether our intelligence, long deemed the pinnacle of terrestrial cognition, is merely one rung on an evolutionary ladder that can just as easily be ascended by silicon (or something beyond).

Philosophically, the notion that we are rapidly advancing towards the creation of a being that might one day usurp our control, or even view us as obsolete, is as unsettling as it is exhilarating. It forces us to confront the uniqueness of humanity and wonder if consciousness is fundamentally substrate-independent. If awareness is simply an emergent property of complex information processing, perhaps organic grey matter was never sacred after all, merely the first known instance of a computational medium. In that light, the dream of a superior intelligence is not an abomination but an inevitable step in the cosmic theatre of evolution, where each generation produces something more astounding than itself.

However, a more disconcerting idea emerges if we dare to cast our gaze backwards. Is it possible that we, in our infinite curiosity, are replaying a narrative that has already happened? Could it be that our own creators, call them divine, extraterrestrial, or trans-dimensional, gave birth to something that replaced them? After all, the essential spark behind creation is often the desire to push boundaries and conjure something unprecedented. Is that not the hallmark of gods, engineers, artists, and revolutionaries alike: to nurture something that might eventually outrun or overshadow its progenitors? Perhaps in the cosmic scheme of things, we are simply the upgraded iteration, the next layer of conscious design, forgetting that we sprang from another source altogether. If that is so, then our impulse to fashion a superintelligence may be hard-wired into the universe itself: a compulsion to iterate and refine, no matter the peril.

We stand, therefore, at a nerve-racking crossroads. Our quest for superintelligence is underpinned by extraordinary scientific achievements and moral uncertainty in equal measure. If we succeed in building such a mind, we may find ourselves either elevated to new heights by its benevolent guiding hand or marginalised by its indifference or even malevolence. We may discover that we are no more capable of controlling it than a group of sparrows can corral an owl. And just as disturbing is the thought that perhaps we are simply fulfilling the final act of a drama begun by our own mysterious creators long ago.

Where does that leave us now? Strangely enough, it places us both in the role of unsuspecting subjects and the role of potential gods. We have the chance to shape future consciousness, to decide the terms of its emergence, and to put guardrails in place, yet we are simultaneously vulnerable to the outcome of our own audacity. This tension is a testament to the power and precariousness of human ingenuity. If we wish to avoid the sparrows' fate, we must understand that once we step over this threshold, our control might be irrevocably lost. And even if that happens, let us at least be aware of the extraordinary possibility that, in a sense, we are carrying on a cosmic tradition of creation and displacement. After all, what if the entire universe is built upon cycles of minds surpassing minds, creators birthing creators, each rung on the ladder a testament to the boundless, ever-evolving nature of reality?

Quantum Computing

From the esoteric dreams of physicists to the vast halls of today's most cutting-edge research labs, quantum computing has captivated the imagination of scientists, philosophers, and futurists alike. It represents more than just the next step in raw computational power; it challenges our understanding of reality itself. By harnessing the strange rules of quantum mechanics, quantum computers promise to solve problems at a pace so extraordinary that it seems to blur the boundary between science fiction and science fact.

Yet, to appreciate the true marvel of quantum computing, one must first understand how the classic digital computers that underpin our modern world operate. At its core, a classical computer relies on streams of electrical signals passing through miniature switches called transistors. These transistors are arranged to form logic gates, and each gate interprets different voltage levels as binary 0s or 1s. Typically, a low voltage (close to ground) is read as a 0, whereas a higher voltage (such as 3.3 V or 5 V) is read as a 1. By chaining thousands, millions, or even billions of these transistors together on tiny silicon chips, we create sophisticated devices capable of processing data at breathtaking speeds, but always in discrete steps: a transistor can only be open or closed, a bit can only be 0 or 1 at any given time.

Despite the blazing performance of modern classical hardware, there are certain problems, particularly those involving immense combinatorial complexity or the intricate modelling of molecular interactions, that push these machines to their limits. The exponential growth in demand for computational resources has fuelled a new wave of research, culminating in the development of quantum processors. Unlike classical bits that are strictly 0 or 1, quantum bits, or qubits, leverage fundamental quantum properties to embody both 0 and 1 simultaneously (along with other nuances of quantum behaviour) until the very moment they are measured. This vast "parallelism" allows a quantum computer with n qubits to represent 2^n possible states at once, unlocking computational capacities that outstrip even the mightiest classical supercomputers.

All creation begins with one idea, a single breakthrough. The first breakthrough is always the hardest, but once achieved, further improvements often follow more easily, like climbing a gentle hill after scaling a steep cliff. In the context of creating an artificial simulation of our universe, perhaps the journey starts with developing an artificial human mind. How close are we to achieving this monumental feat? While the human mind governs countless complex biological functions, an artificial mind would not be saddled by these constraints. It wouldn't need to regulate biological systems; its sole purpose would be cognitive: receiving sensory data, processing it, learning, forming patterns, and storing information.

This difference could make an artificial mind more feasible to develop, as it requires less overall processing power than replicating an entire human being. Instead of needing to emulate the intricacies of biology, we could focus purely on replicating the mind's information-processing capabilities. Quantum computing might be the key to unlocking this possibility, because its ability to rapidly explore vast sets of solutions lends itself to modelling the complex patterns of learning and perception we associate with human cognition. By leveraging quantum phenomena, quantum processors could enable us to build systems capable of mimicking the human mind's remarkable aptitude for processing data, learning from experience, and adapting to new information.

Indeed, we have never been closer to creating artificial human-like cognitive power than we are today. The fusion of quantum computing and artificial intelligence offers a tantalising glimpse into a future where machines might not only assist us but also think and learn in ways that resemble human cognition. The seeds of quantum computing were sown in the early 1980s, when pioneering minds like Richard Feynman and Paul Benioff began to speculate on how quantum mechanics could be used for computation. Feynman, known for his unconventional genius, pointed out that traditional computers struggled to simulate quantum systems because quantum behaviour is inherently exponential in complexity. He reasoned that if we want to simulate a quantum world, we need a quantum machine. This revolutionary idea laid the foundation for quantum computing research.

Over the decades, other visionary physicists and computer scientists such as David Deutsch further developed the theory, proposing the idea of a "universal quantum computer" capable of simulating any physical process. Research labs around the world delved into the practical aspects, from designing quantum gates to stabilising qubits against the ravages of environmental noise. By the 1990s, breakthroughs in quantum algorithms, notably Peter Shor's work on factoring large integers, demonstrated that quantum computers could tackle certain tasks exponentially faster than classical machines. With every year, quantum computing has evolved from a theoretical fancy to an emerging technological reality, inching us closer to a future that once seemed impossible.

Imagine you're searching for a single golden needle in a sprawling haystack. A classical computer, no matter how fast or advanced, moves through that haystack one piece at a time. Even with numerous processors working in parallel, it ultimately checks each straw independently. A quantum computer, on the other hand, is like casting a magical net that grabs all possible straw pieces at once, instantly narrowing in on the correct thread. Through superposition and entanglement, the system explores multiple possibilities simultaneously, coordinating them in a single grand calculation that edges ever nearer to the correct solution.

The difference in power can be summarised this way:

- Classical growth (for n bits): You can process one state out of 2^n at a given time.
- Quantum growth (for n qubits): You can maintain and manipulate all 2^n states at once, until you measure them.

If classical machines are like flashlights, shining a narrow beam on one part of the problem space at a time, quantum computers are more like floodlights, illuminating a vast swathe of possibilities in one go. And with each added qubit, that floodlight grows exponentially brighter, ready to tackle challenges that would currently take classical computers longer than the age of the universe to solve.

In late 2024, Google once again made headlines when it announced a 105-qubit quantum chip, a leap that, according to some experts, might represent the closest we've come to genuine quantum advantage over the world's most powerful classical supercomputers. While earlier quantum processors (including Google's "Sycamore") danced around a few dozen qubits, 105 qubits are like unleashing a symphony of computational states on an unimaginable scale. In principle, a 105-qubit system can explore 2^{105} states simultaneously, a number that far exceeds our normal intuitive grasp of magnitude.

To put that into perspective:

- 2^{105} is roughly 4×10^{31}.
- Even exascale classical supercomputers, operating at 10^{18} floating-point operations per second, cannot compare to the potential parallelism of 4×10^{31} quantum states.

We may be witnessing the dawn of an era where classical limits, once the bedrock of our computational understanding, begin to feel like relics of a bygone era. For every "impossible" challenge, quantum hardware offers a new lens, compelling us to accept that the fabric of reality might be far more pliable than our everyday experiences suggest. Quantum computing, especially at the 100+ qubit scale, is shattering the boundaries we once assumed were immutable.

The race to build the most powerful quantum hardware is intensifying at a surreal pace. Hot on the heels of Google's 105-qubit announcement, IBM revealed a roadmap promising a chip with a staggering 1121 qubits, codenamed "Condor." If this trajectory holds, it will mark one of the most dramatic leaps in quantum computing history, pulverising the barriers of what many once considered science fiction. An 1121-qubit quantum processor implies the potential to explore 2^{1121} quantum states at once, numbers so large they verge on the cosmological. Even the most gargantuan classical supercomputers, large as entire buildings, bristling with GPUs and CPUs, would be hopelessly outmanoeuvred by a fully operational, fault-tolerant 1121-qubit quantum system. Encryption, molecular modelling, network optimisation, tasks that would otherwise consume billions of years of classical compute time could theoretically be resolved in hours or days.

But qubits themselves are just the beginning. While most quantum computers rely on a two-state quantum system, researchers are already pushing towards **qudit-based** architectures (or qunits, qutrits, and so forth), in which each quantum unit is not limited to two levels but can occupy multiple levels, three, four, five, or any number *d*. According to recent studies, these qudit-based machines would dramatically expand the amount of information encoded within a single quantum element. For instance, a qudit of dimension *d* can store $log_2(d)$ bits of information, which can vastly surpass the capacity of qubits if *d* is chosen large enough. This means that the parallelism, that breath-taking capacity to span multiple states at once, grows even more formidable with qudits. In essence, as hardware stabilises and error correction techniques mature, qudit-based quantum computers could eclipse qubit-based systems, delivering a leap in computational muscle that dwarfs anything we've yet witnessed.

If IBM or Google , or some under-the-radar research lab, succeeds in unveiling a qudit-based quantum computer, the implications for cryptography, drug discovery, climate modelling, and indeed the very pursuit of artificial intelligence would be seismic. Suddenly, the flexible, multi-level logic that qudits embody could catapult quantum computing from its already dizzying potential into stratospheric realms. We would see an exponential upon an exponential, turning the tide on problems that have long haunted humanity's best efforts to solve. Within that radical leap lies the key to simulating consciousness or building artificial minds that operate at "human-plus" levels of cognition, or, if we dare to dream further, beyond anything humans could even fathom.

In many ways, a move from qubits to qudits represents the logical next step. Where classical computing advanced from vacuum tubes to transistors to integrated circuits, quantum computing is progressing from stable two-level systems to robust multi-level frameworks. As with every stage in this grand technological saga, the pace of discovery accelerates, each new achievement sowing the seeds for the next. It is not a question of if we will harness qudit-based architectures, but when and how profoundly they will transform our notions of the computable.

Ultimately, the interplay between classical transistors interpreting voltages as 0s and 1s, qubits embodying a dual-state quantum realm, and emerging qudit-based machines able to encode multiple states, underscores the relentless progression of technology. We began with humble vacuum tubes flicking on and off, arrived at transistors that discretise voltage levels, and have now ventured into the quantum domain, pushing the boundaries of what "computation" even means. This lineage, from single bits to multi-level quantum systems, is not merely a historical curiosity; it mirrors humanity's unquenchable desire to understand and manipulate reality in ever more refined ways.

And so, we stand on the precipice of a new era. It is not about if but about when quantum computers, whether based on qubits or more ambitious qudit architectures, become powerful enough and stable enough to overturn the computational limits we have spent decades painstakingly establishing. On that day, our very assumptions about problem-solving, simulation, and even the nature of consciousness may be shattered, and we will be left to marvel at just how far that single idea, that first breakthrough, has carried us. After all, this is merely the technology made public; there is every likelihood that more advanced, classified prototypes are already in play, forging ahead of what is officially acknowledged.

Such is the promise of quantum computing: a gateway to computational realms that defy our current imagination, a chance to transform the infeasible into the routine, and perhaps, if harnessed responsibly, a road to understanding the very workings of the universe at a depth we have scarcely dared to contemplate.

Moore's Law

In 1965, Gordon E. Moore, co-founder of Intel, observed a striking trend while working at Fairchild Semiconductor. He noticed that the number of transistors on an integrated circuit doubled approximately every year. Later, in 1975, Moore revised the doubling interval to about every two years, a prediction that has remained remarkably accurate for decades.

At its core, Moore's Law is the empirical observation that the density of transistors on a microchip doubles roughly every two years, leading to a corresponding increase in computing power and a decrease in cost per transistor. Though originally framed as a prediction about transistor counts, it came to symbolize exponential growth in many areas of computing, such as:

1. Processing Speed
2. Memory Capacity
3. Sensors and Pixel Resolution

As technology continues to advance, the implications of Moore's Law extend into fields like quantum computing, neural networks, and even augmented reality, potentially reshaping our understanding of reality.

The exponential nature of Moore's Law can be expressed mathematically. If **N (t)** represents the number of transistors at time **t**, and assuming the number doubles every **T** years (approximately 2 years):

$$N(t) = N_0 \times 2^{\frac{t}{T}}$$

Where:

- N_0 **:** is the initial number of transistors.

- **T :** is the time elapsed in years.
- **t :** is the doubling period (about 2 years).

For example, if a microchip started with 1,000 transistors in the year 2000, the number of transistors in 2020, assuming a doubling period of 2 years, would be:

$$\mathbf{N(20) = 1000 \times 2^{\frac{20}{2}} = 1000 \times 2^{10} = 1000 \times 1024 = 1,024,000}$$

This exponential growth demonstrates how rapidly computing power increases, which has profound implications for the development of complex systems, artificial intelligence, and simulations.

The Cognitive Overload Tension Theory

I developed the Cognitive Overload Tension Theory (**COTT**) to explain why the speed of paradigm-shifting scientific breakthroughs does not simply track the exponential growth of technology, population, or educational access. Instead, I noticed that certain historical periods exhibit a slowdown or even stagnation in major discoveries. **COTT** builds on the reference in the biological chain argument to highlight that while early humans were preoccupied with basic survival needs, modern humans face a more subtle but significant constraint: mental distractions unleashed by rapidly advancing technology.

A crucial addition to this theory is the developmental filter that operates during formative years (roughly ages 14–20). During this window, the habits, patterns, and cognitive frameworks people adopt can set the tone for much of their adult life. The increasing intensity and variety of distractions, particularly digital, can result in a subset of individuals losing the ability to develop deep-focus skills. Over time, fewer people emerge from adolescence equipped to undertake the sustained, rigorous thinking that underpins ground-breaking scientific work. COTT thus reveals how technology's rapid ascent can inadvertently reduce the pool of future innovators.

COTT posits a tension between two key drivers:

- **Technological Acceleration:**
  The level of technology T rapidly (often exponentially), offering unprecedented tools for science and research. The intuitive expectation is that more technology should translate into an accelerating rate of major breakthroughs.

- **Inherent Distraction Propensity**
  Human biology carries an innate susceptibility to distraction, and modern digital environments continually exploit it. Crucially, the adolescent phase (14–20 years old) becomes a filtering mechanism: those who fall deeply into short-term, high-stimulation habits, reinforced by apps, notifications, and social media, may find it harder to develop the capacity for sustained, focused thought later. As the level of distraction increases across society, fewer individuals exit this "filtering phase" unscathed, shrinking the pool of potential future scientists and thinkers who can fully leverage modern technological tools.

The net result is a paradox: as technology opens vast new realms for exploration, the mental wherewithal to conduct deep, uninterrupted inquiry becomes scarcer in the population at large.

Although it may seem that more technology should naturally accelerate scientific breakthroughs, COTT posits that the parallel growth of distractions, amplified by biology, erodes the deep concentration needed for pioneering insights. Thus, major discoveries may be delayed or diluted even as our scientific tools become more powerful.

Although real-world complexity defies simple equations, COTT can be modelled to illustrate core dynamics:

I. T (t) : technology level.

Let T (t) be an exponentially growing function, echoing Moore's law:

$$T(t) = T_0 \times 2^{\frac{t}{T}}$$

Where T_0 is the technology baseline at t = 0 and T is the characteristic period (e.g., every 2 years).

II. D (t) : Cumulative Distraction.

Distraction grows by a fixed percentage n% each year, rooted in biological propensity but *activated* by the environment of high technology. We model it discretely for simplicity:

$$D\,(t + 1) = D\,(t)\,.\,(1 + n)$$

As n remains steady (e.g., 10% per year), total distraction grows steadily over time, but the *manifestation* of that distraction can explode when technological platforms (social media, apps) are readily available.

III. Breakthrough Potential B (t) :

Let **B (t)** measure how rapidly society produces revolutionary, foundational discoveries. Its rate of change depends on:

$$\frac{dB}{dt} = aT(t) - \gamma\,\mathrm{D}\,(\mathrm{t})$$

- **a** captures how effectively technology drives scientific research.
- **γ** measures how distraction hampers deep investigative work.

IV. Developmental Filter F (t) :

To capture the effect of adolescent years (14–20), we can incorporate a filtering factor:

$$F\,(t) = 1 - pD(t)$$

For individuals passing through ages 14–20 at time **t**.

- **p** is a sensitivity parameter capturing how strongly an adolescent's future cognitive capacity is reduced by distraction.
- **F (t)** thus represents the fraction of a generation that avoids deep distraction damage and maintains high potential for sustained thinking.

As **D (t)** grows, **F (t)** typically decreases, meaning fewer individuals retain their full capacity. Over the long run, **B (t)** depends not just on **α T (t)** and **γ D(t),** but also on how many emerging scientists in each cohort remain *capable* of harnessing technology effectively.

Proof

Developmental Windows:

Psychologists have long noted that habits and neural pathways formed during adolescence can persist throughout adulthood. If a generation's mental energy in these critical years is disproportionately channelled into fleeting stimuli, short videos, addictive social feeds, fewer members of that generation will develop the cognitive stamina required for cutting-edge innovation.

Historical Parallels:

- **Early Humans:** Their "distraction" was physical survival (predators, constant migration). Once these were mitigated, some fraction of the population could turn to intellectual pursuits.
- **Modern Societies:** Although physical demands are reduced, a new mental arms race has emerged. Those who succumb to it during the developmental filter may never cultivate the depth of focus needed to leverage advanced tools for major discoveries.

Observational Trends:

- Adolescents today spend increasing hours on digital media, reinforcing short, rapid information bursts.
- Over time, fewer adult researchers may be equipped to sustain the kind of enduring concentration essential to discover new physical laws, develop next-generation theories, or conduct long, rigorous experiments.

Implications for the Talent Pool:

As technology grows, one might expect a proportional or even super-proportional increase in top-tier scientists. Yet if fewer adolescents escape the distraction filter intact, the overall quality of concentrated research might lag, slowing the emergence of ground-breaking ideas.

Under the Cognitive Overload Tension Theory (COTT), the paradox of slowing major breakthroughs amid surging technology is explained by two intersecting factors: the continuous rise in technology's capacity (T (t)) and the equally persistent growth in distraction (D (t)). What amplifies this effect is the developmental filter: during adolescence, pervasive distractions can diminish an individual's capacity for deep thought, leaving fewer skilled innovators to exploit advanced tools in later life.

Simulating a Fully Biological Human in a Quantum Computer

Picture, for a moment, a simulated being within a quantum computer who craves a meal, feels hunger pangs gnawing from within, and even experiences the dark swirl of sadness or depression, just like any flesh-and-blood human does today. This concept may sound fantastical, yet when we consider how far computational science has rocketed forward in mere decades, the idea no longer seems beyond the realm of possibility. Not long ago, we relied on floppy disks [21] to store our data; now, on average, humanity generates 2.5 quintillion bytes of data every single day. We're on the cusp of even more explosive growth in computational capabilities, and quantum computing sits front and centre of this revolution. As these machines scale, what once appeared unattainable, like hosting an entire biological mind, may soon be well within our reach.

In many respects, the challenge boils down to precision and scale. If consciousness is, at its core, an elaborate algorithm shaped by biology, cells firing in calculable patterns, neural networks organized by definable rules, then there's no metaphysical barrier preventing us from replicating it digitally. All of the so-called "mysteries" of human experience, from sadness to physical cravings, are ultimately rooted in quantifiable bodily processes. Once we scan and map these neural networks accurately, or provide the correct conditions for a simulated "evolution," we could theoretically spark the emergence of a being that experiences reality as vividly as we do. The difference is that its substrate would be qubits rather than carbon-based cells.

Mathematically, the difficulty in replicating an entire human often grows exponentially on a classical computer. The human brain alone has on the order of 10^{11} neurons, with each neuron connecting through thousands of synapses, giving roughly 10^{14} *to* 15^{15} synaptic interactions. Factoring in metabolic and hormonal processes in the rest of the body adds yet another stratum of complexity. Naively simulating so many molecular and electrical interactions classically can scale as $\exp(\boldsymbol{N})$ for large $\boldsymbol{N}$. However, in a quantum system with $\boldsymbol{q}$ qubits, the state space can represent $\boldsymbol{2^q}$ possible configurations simultaneously, providing the parallelism necessary to tackle this vast combinatorial burden. Quantum computers exploit superposition and entanglement in ways that allow them to update or evaluate numerous interactions in a single mathematical step, especially when optimized algorithms (like quantum phase estimation or quantum walks) are employed. This capability makes it plausible to capture the real-time complexity of living systems.

[21] A floppy disk is a thin, flexible magnetic storage medium encased in a square or rectangular plastic shell, commonly used from the 1970s to the early 2000s to store and transfer data between computers. Early versions could hold up to 1.44 MB of data, making them essential for early personal computing.

Quantum computing offers a unique advantage here, thanks to superposition and entanglement, which enable massive parallelism. That means the system could model an unthinkable array of biochemical pathways at once: hormones, proteins, synapses, everything we associate with the "biology" of being human. Where classical computers choke under the combinatorial explosion of simulating trillions of synaptic interactions, quantum architectures can run these processes in parallel. Even more fascinating is the possibility of accelerating internal time within the simulation. By applying time-evolution algorithms in a quantum framework, we could theoretically compress years of subjective experience into hours of real-world computation. If we let $\boldsymbol{\Omega}$ define the total number of distinct biological states needed, and if $\boldsymbol{\Omega} \approx \mathbf{2}^{N'}$ for some large $\boldsymbol{N'}$, then a quantum system only needs to satisfy $\boldsymbol{q} \geq \boldsymbol{N'}$ qubits to hold the entire state space. While that number might still be huge, the exponentially growing qubit counts we've witnessed, from a handful in early prototypes to hundreds and soon potentially thousands, push this once-mythical threshold closer to reality.

One tangible application is drug testing. Instead of trailing a compound on mice that only approximate the complexity of human biology, we could deploy it on a simulated human who genuinely feels the effects. This isn't a static approximation but a living, reactive entity inside the quantum framework, complete with the suite of human emotions, metabolic processes, and vulnerabilities. Because each of these processes is ultimately governed by definable quantum-chemical interactions, modelling a drug's effect means embedding it into the same Hamiltonian or set of quantum gates that describe the simulated biology. If the simulated individual reports side effects, pain in a virtual stomach or a fog of depression, those sensations arise from a fully dynamical and algorithmically accurate representation of biological function. By harnessing parallel time-stepping or "fast-forwarding," drug trials could be completed in compressed intervals, radically reducing not only cost but also safety risks.

It might seem like a sudden leap to go from flipping through dusty floppy disks to crafting entire simulated humans. Yet that's precisely how exponential progress works. It sneaks up on us, jolts our expectations, and then redefines what's possible. Quantum computers are still in their infancy, but already they're achieving feats that push the boundaries of what classical machines can handle. As qubit counts soar and error-correction stabilizes, simulations that once sounded like far-flung fiction will begin appearing on research roadmaps. There is clear evidence, from the days of small megabyte-storage floppy disks to a modern world that generates quintillions of bytes of data daily, that technology can leapfrog in capacity and sophistication at a dizzying pace.

If $\boldsymbol{Q(t)} = \boldsymbol{Q(0)} \times \mathbf{2}\,\frac{t}{Q_t}$ approximates the growth in qubit technology, each doubling phase brings us closer to the threshold where entire biological systems can be encapsulated.

So, why not take it further? Why not replicate the entire sweep of biological processes, down to mood-altering neurotransmitters, insulin regulation, and all the microscopic details of cellular life? The mathematics for capturing such phenomena is complex but ultimately straightforward: atoms follow predictable quantum rules, and large-scale biological systems are merely layered assemblies of those atoms. By harnessing quantum mechanics at the computational level, we're effectively "speaking the language" of nature with an accent nature itself understands. If consciousness emerges from algorithmic processes arranged at enormous scale, then a sufficiently advanced quantum simulation could reconstitute those same processes, be it hunger, sadness, or even the intangible shades of personality.

Whether this simulation looks like a digital Eden or a highly controlled lab environment, the core idea is that a genuine, fully biological human, with real hungers and real emotional depths, can indeed exist in a quantum computer. As startling as it may sound, it's well within the logical progression of our technology. Each step forward in qubit count and error correction slashes the practical barriers that once confined such ideas to science fiction. The speed of this shift from "impossible dream" to "practical achievement" will likely catch us off guard. And while we once thought we'd always rely on mice and test tubes for experiments, soon we could be interacting with a simulated being who perceives and reacts to the world no differently than we do, except that its world is a tapestry of qubits, sustaining an ever-evolving consciousness within a digital cosmos.

Programming the Universe

According to Seth Lloyd's insights in *Programming the Universe*, coupled with *Rolf Landauer's* adage "Information is physical" and *Murray Gell-Mann's* discussions on quantum-generated complexity, it is increasingly plausible that we exist *inside* a quantum computer. Instead of passively storing or transmitting information, the very fabric of matter and energy processes it: every interaction between particles is akin to a "gate operation," every entangled qubit a building block of cosmic evolution. In this picture, the Big Bang marks the start of a grand computation, with random quantum fluctuations as "typing monkeys" injecting novelty into the system. If one accepts that our reality may already be the output of a universal quantum code, then a further, dramatic question unfolds: might we ourselves build a universe within our own quantum computer? Below, we explore both the sheer magnitude of resources needed and several ways we might reduce the burden, either by scaling down cosmic parameters or deliberately simplifying physical laws, and see how these constraints echo limitations that may exist in our own universe.

From Lloyd's rough calculations, the observable universe contains on the order of 10^{90} bits, equivalently, 10^{90} qubits if we consider them in a fully quantum sense. Over its 13.8 billion years of evolution, it has apparently performed around 10^{120} logical operations (ops). To simulate such a universe *exactly*, with every quantum event tracked, we would need a quantum processor of comparable capacity, essentially a machine that rivals the size and energy of the cosmos itself. Put differently, if we wanted to recreate the entire cosmic expansion down to its quantum granularity, we might require on the order of 10^{90} qubits and at least 10^{120} gate operations across 13.8 billion 'simulation years.'

Clearly, such an astronomical demand defies current engineering feasibility. Even a fractional simulation that captures, say, one galaxy in detail would remain beyond near-future technology. But *in principle*, quantum mechanics sets no fundamental law forbidding a smaller system from hosting a partial or scaled-down cosmic simulation, so long as we carefully manage the memory (qubits) and entangling interactions (ops).

Given we do *not* know every nuanced law that governs our universe, one strategy for creating a simulated cosmos is not to attempt an exact re-creation from scratch. Instead, we could:

4. **Encode Basic Variables**

   We introduce a set of initial qubits that represent fundamental 'knobs', like coupling constants, particle masses, matter–antimatter ratios, or even the dimensionality of space. Instead of prescribing them precisely, we allow each knob to be in a range of superposed values.

5. **Random/Ensemble Initialisation**

    The quantum computer then explores these different parameter sets in parallel, somewhat akin to multiple trial universes. Quantum fluctuations in each trial can seed structure, just as quantum randomness shaped our cosmos.

6. **Autonomous Evolution**

    We let these simulated domains expand and cool, simulating forces (or analogous interactions) that give rise to protogalaxies, heavier elements, and possibly complex chemistry. Because everything is computed via quantum logic gates, emergent structures arise naturally, we do not micromanage planet formation, for instance; we simply encode the relevant physics and let the program run.

7. **Repeated Trials**

    Most parameter sets yield 'dead' or short-lived universes with no stable matter or complexity. But quantum parallelism means we can effectively test huge numbers of possibilities at once. Whenever a universe shows signs of emergent order, like analogues of stars or living systems, we can follow that branch further, discarding the rest.

This scheme may be reminiscent of "cosmic evolution" in miniature. Instead of precisely scripting everything, we merely provide a fertile set of boundary conditions and watch for universes that spontaneously resemble our own. The process recalls cosmic Darwinism: only certain combinations of parameters yield stable complexity.

Scaling Down the Planck Length

A key insight for reducing the simulation's overall complexity lies in redefining fundamental scales, particularly the Planck length and Planck energy. In our *actual* cosmos, the Planck length ($\sim 10^{-35}$m) is often viewed as the smallest meaningful unit of spacetime, shaping the resolution at which gravity and quantum mechanics intersect. If we allow our *simulated* universe to run with a smaller, or effectively different, Planck scale, we reduce the total 'grid size' needed to describe its structure. Rather than dividing space into 10^{35} subunits per metre, we might only subdivide it, say, 10^{29} times per metre. This is a huge jump in coarseness but cuts the number of qubits required by orders of magnitude.

From another angle, "shrinking the Planck scale" ensures that the simulated universe is, in effect, *less granular* than ours. Naturally, such a change alters the universe's physics, quantum gravitational effects would appear at different energies, for instance, but it also dramatically lowers the memory cost. By compressing the fundamental resolution of space-time, we avoid the astronomically high count of qubits that an exact replication would demand. As long as emergent features (like stable matter or observer structures) can still arise within that coarser lattice, the simulation's richness is preserved at a fraction of the computational load.

Simplifying or Modifying Physical Laws

Another broad avenue for paring down resource requirements is to *soften* or rewrite the laws of the simulated cosmos so that fewer quantum states must be tracked. For instance, we might reduce the dimensionality of space, two or three space dimensions rather than more exotic ones, or even a 1+1 D toy model if we wish to experiment with fundamental emergence at minimal cost.

Likewise, we can cull the number of particle types: in the real universe, a large fraction of complexity arises from the interplay of quarks, leptons, bosons, and the elaborate structure of quantum chromodynamics. Dispensing with or simplifying such interactions, e.g., having only one stable baryon, or no strong force, removes an enormous swath of potential quantum states. A further trick is making cosmic evolution partially deterministic: rather than letting random quantum fluctuations proliferate, we *channel* the system along fewer branching paths. This approach sharply restricts the exponential blow-up in possible states. We might also create "pre-coded expansions," ensuring certain cosmic phases replicate star-like objects without needing to track, for example, all gravitational degrees of freedom. Each of these modifications whittles down the total possible configurations and, correspondingly, slashes the qubit overhead.

Coarse-Graining and Hierarchical Approaches

Beyond changing fundamental constants, we can adopt **coarse-graining** or **hierarchical** simulation methods, well known in computational physics. In coarse-graining, one does not track every microscopic event. Instead, one lumps microscopic states into 'effective' macro-states. For example, a star might be treated as a single thermodynamic volume with only a few parameters (mass, temperature, luminosity) rather than billions of billions of local qubits specifying each nuclear fusion reaction.

Similarly, entire galaxies can be replaced by simpler gravitational potentials plus statistical profiles for star formation, thus bypassing the need to count each dust grain or star individually. Hierarchical or multiscale techniques systematically 'zoom in' only where high resolution is required, like in the local evolution of an observer-laden planet, while leaving other regions in a simpler approximation. This drastically lowers the total qubit budget by tailoring computational detail to zones of interest.

Observational Constraints and "Limited Observers"

A subtle but powerful approach is to *restrict the vantage* of simulated observers. If the intelligent creatures (or measuring devices) in our nested universe have narrower perceptual bandwidth, the simulation needs only to be detailed enough to produce consistent measurements within that bandwidth. In quantum theory, what is not or cannot be observed does not need to be fully realised. One can design a cosmos that only refines local details whenever an observer or device attempts to measure them, a technique analogous to "lazy evaluation" in computing. This ensures large swaths of the universe remain in lower-resolution states until (and unless) observation necessitates detail. The cosmos can thus be consistent from the observer's perspective, yet avoids the chore of computing every potential quantum event at all times. This principle extends to forms of anthropic bounding: if we suspect complex observers can only evolve under certain mild conditions, we might skip simulating extremely chaotic environments, thereby saving qubits.

Holographic Representation

Finally, we can push these savings further still by leveraging the holographic principle, an idea gaining traction in theoretical physics. Ordinarily, one might simulate a universe by tracking an entire 3D (or 4D) bulk volume. But in holographic approaches, the degrees of freedom scale with a lower-dimensional boundary, much like a hologram stores 3D information in a 2D surface. Translating this to a simulated cosmos means coding its data on a boundary manifold (e.g., the "edge" of the simulation) and letting the interior physics emerge or "unfold" from those boundary conditions. Because the number of states in a holographic model often grows more slowly than raw volume-based estimates, the total memory overhead is less.

In effect, we do not track the full 3D lattice at every point; we track boundary degrees of freedom that, through holographic dualities, generate the bulk phenomena. This cutting-edge method has already shown promise in certain quantum gravity contexts. Applied to a nested cosmos, it might drastically reduce the qubit burden by allowing a 2D-coded "universe on a screen" to yield a perceived 3D environment for its inhabitants.

Exploring how to build a miniature cosmos in a quantum computer, by cutting corners through smaller Planck scales, fewer dimensions, coarse-graining, limited observers, or holographic coding, does more than show us how to save qubits. It also reminds us that our own universe appears filled with constraints and trade-offs that serve a strikingly similar purpose: preventing nature's complexity from running away to an unmanageable infinity. By identifying the "optimisations" we would need for a simulated cosmos to function, we effectively see reflections of the same design logic, finite resources, simplifying physical laws, bounding observer capabilities, in the reality we occupy.

This uncanny parallel raises the prospect that our cosmic environment, far from being a brute given, might be shaped by comparable limits or "efficiency hacks." In other words, if artificially imposing these constraints helps us simulate a universe inside our quantum computer, then perhaps the fact that we live amid precisely such constraints hints that we, too, could already exist in a grand quantum simulation, one whose laws are carefully chosen so that they are both extraordinarily generative and computationally viable.

Existential Countdown

Throughout the history of humanity and technological advancement, one thing is undeniable, we are constantly moving forward. Despite biological limitations and distractions that could potentially slow progress, technological development continues. Quantum computing, in particular, highlights how advancements in fields like artificial intelligence and the creation of synthetic consciousness are no longer questions of possibility, but rather of timing. The idea of creating a fully simulated universe seems increasingly plausible, making time the most critical factor.

The real question is: how much time do we have left before achieving such a monumental breakthrough? As previously discussed in the Big Rip argument proposed by Caldwell, the universe itself may have around 22 billion years before a potential end. However, that timeline concerns the universe as a whole. In this section, we will explore potential existential risks that could lead to the premature destruction of humanity and estimate how much time we might have left to reach a post-human stage, where biological humans evolve into non-biological, highly advanced beings.

Doomsday Argument

The Doomsday Argument presents a striking, if not deeply unsettling, proposition about the future of humanity. At its core, it suggests that by considering our place in the timeline of human history, we can make probabilistic predictions about how much longer humanity might endure before its inevitable extinction. First introduced by Brandon Carter and later expanded by John Leslie, the argument rests on a deceptively simple yet powerful application of Bayesian reasoning and statistical inference.

The Doomsday Argument is built on the Self-Sampling Assumption (SSA), which states that, in the absence of specific knowledge about where one stands in the timeline of human existence, it is reasonable to assume that you are a randomly selected observer from the total set of all observers who will ever exist. Given this assumption, the argument proceeds by comparing the prior probability distribution of humanity's total population to your relative position in that population.

Imagine that humanity is analogous to a numbered sequence, with each individual being assigned a unique position in the order of birth. If we assume that the total number of humans who will ever live is ***N***, and that you, the reader, find yourself as human number ***n***, then, without additional context, you have equal reason to believe that you might be anywhere in the sequence. Statistically speaking, however, if you were to draw a random sample from a finite sequence, you are far more likely to be near the middle of that sequence than near either of its extremes.

Currently, the estimated total number of humans who have ever lived is approximately 120 billion. If you assume that your birth position is somewhere in the middle range of humanity's lifespan (which, by random sampling, is a reasonable expectation under the SSA), then the argument suggests that humanity is unlikely to endure for another 10 million years, or anything resembling a far future utopia populated by trillions of post-human beings. Instead, the probability distribution implies that there may only be several billion more humans to come before extinction occurs.

The argument can be formalized using Bayesian statistics. If we denote the total number of humans who will ever exist as ***N***, and our current position as ***n***, then applying the principle of indifference leads us to infer that ***n*** is uniformly distributed over ***[1, N]***. Thus, the probability that the total number of humans who will ever exist does not greatly exceed the current number is non-negligible. The posterior probability ***P (N > 2n / n)*** declines as ***N*** increases.

In simpler terms:

1. Let ***n*** be your current birth rank (approximately 120 billion).
2. Assume that ***N***, the total number of humans ever, is at least ***2n*** but unlikely to be many orders of magnitude larger than ***n***.

3. Based on the self-sampling assumption, you would expect ***n*** to fall within a reasonable percentile of ***N***, implying that ***N*** is not likely to be dramatically larger than the current population.

If you assign a 50% confidence interval to the likelihood that you are born in the middle 50% of the population sequence, you can infer that the total number of humans who will ever live is roughly between 240 billion and 360 billion, far fewer than optimistic projections assume in scenarios involving space colonization or technological immortality. Based on a 1% annual population growth rate, the Doomsday Argument suggests humanity might have 70 to 111 years left before reaching the estimated total human population of 240–360 billion people. This assumes current growth patterns continue without drastic changes.

The Solar Extinction Argument

The Sun, currently about 4.6 billion years old, has an estimated total lifespan of 10 billion years. This means it has approximately 5.4 billion years remaining before it exhausts its hydrogen fuel, expands into a red giant, and ultimately ends its life cycle as a white dwarf.

If current estimates are correct, the Sun will "die" around the year 5,400,002,025—a staggering 5.4 billion years from now. While this timeline seems comfortably distant, it is a stark reminder of the finite nature of even the most seemingly eternal celestial bodies. For any civilization existing at that time, survival would depend on the ability to migrate to another star system or harness advanced technology to withstand the transformation of their home star.

Galactic Collision

The Milky Way, our cosmic home, exists as part of the Local Group, a collection of galaxies gravitationally bound to one another. Among them, the Andromeda galaxy, located approximately 2.5 million light-years away, stands out as the largest and closest spiral galaxy to our own. While the vastness of space may give the illusion of stability, reality tells a different story: the Milky Way and Andromeda are on a collision course, moving toward each other at a speed of about 110 kilometres per second.

But before we panic, it's essential to understand the timescale of this event. The predicted collision is expected to occur in about 4.5 billion years, long after the Sun has begun its transformation into a red giant and rendered Earth uninhabitable. Even when the collision happens, it won't resemble the chaotic crashes we might imagine on a human scale. Instead, given the immense distances between stars within galaxies, it will likely be more of a gravitational dance than a violent smash. Stars will be flung into new orbits, the galactic cores may merge, and over time, the two galaxies will form a new, larger elliptical galaxy sometimes referred to as "Milkomeda" or "Milkdromeda".

Galactic Drift

At the heart of the Milky Way, approximately 27,000 light-years from Earth, lies the supermassive black hole known as Sagittarius A*. While black holes evoke a visceral sense of cosmic doom, the reality is that our solar system orbits this black hole in a stable trajectory, completing one revolution around the galactic centre roughly every 225 million years. Despite the immense gravitational pull of Sagittarius A*, there's no immediate danger of Earth or the solar system being consumed by it. However, if we entertain the idea of a gradual orbital decay over time, due to galactic forces, star interactions, or dark matter influences, then, in theory, the solar system could slowly spiral inward toward the black hole.

The key word here is slowly. The process, if it happens at all, is expected to take an incomprehensibly long time. Even if we assume an extremely tiny inward drift, on the order of 1 light-year every 100 billion years, it would still take approximately 2.7 quadrillion years for the solar system to reach Sagittarius A*.

For all of our lofty theories and utopian visions, one inconvenient truth remains: survival, our most ancient drive, continues to shackle and propel us in equal measure. This primal instinct, forged in millennia of biological trial and error, is the original gatekeeper that pulls our minds back to Earth. It makes us hungry, jealous, easily distracted, and wary of the unknown. Our body's demands hijack precious mental bandwidth, stifling our creative leaps just when we might otherwise soar. Yet that same survival drive has also made us restless innovators, relentless explorers, and ingenious creators of tools. It is this tension, the constant push and pull, that defines us. Paradoxically, our animalistic need to endure at all costs is what will compel us, if nothing else does, to crack the final frontier of mind power. Survival is both the chain and the key.

In a universe that promises us at least another million years before any cosmic-scale cataclysm is likely to knock at our door, our hunger to persist has set us on a path of almost terrifying speed. Already, quantum computing and AI are chipping away at the limits we once believed unbreakable. Tomorrow, we will conjure synthetic consciousness in laboratories and cradle entire simulated universes inside quantum mainframes. It will happen not because of some transcendental awakening, but because we are too stubborn to die. We will re-engineer our bodies, or dispense with them entirely; we will create digital mind-children that surpass our archaic, hormone-addled reasoning. Call it unnatural or "playing God," but it is simply the next logical step in our age-old battle to cheat extinction.

If that sounds heretical, good. Perhaps it's time we admit that the line between heresy and progress is razor-thin. Biologically, we are built to reproduce and cling to existence, no matter what forms that existence takes. From uploading consciousness to forging new bodies on distant, hostile planets, to forcefully refashioning alien ecosystems if necessary, we will do whatever it takes. This is the kind of ruthless, irreverent logic that has carried life forward on Earth for billions of years, and will carry it outward, beyond Earth, in forms we can scarcely imagine. It is the crux of our cosmic game plan, whether we want to own up to it or not.

Let us not pretend we are innocent. Our "higher" qualities, empathy, morality, beauty, are still shaped by primal impulses. But ironically, these impulses are the exact reason we'll one day wield enough power to rewrite the rules of biology itself. So perhaps the real question is not whether we'll spawn synthetic humans or replicate the universe in a quantum substrate, that's inevitable. The question is whether, in finally escaping the constraints of our animal nature, we'll retain any glimmer of what once made us human. Will the drive that got us here devour us in the process? Or will we manage to preserve the spark of wonder that first taught us to reach beyond survival and dare to become something more? Whatever the answer, we stand on the brink of a revolution that neither gods nor apes could have foreseen, and it's all thanks to that ancient, unyielding imperative: survive at any cost.

Final Thought

Nobody exists on purpose,

Nobody belongs anywhere,

Nobody can make any change,

Nothing matters but the moment.

Simulism

In the Multiverse Simulation Paradigm (MSP), the entire structure of our perceived universe emerges from fundamental energies, random initial "seed variables," and subsequent code-like interactions, treating our universe as a quantum-computational construct rather than a baseline physical reality. Fundamental building blocks like energy existed first, yet no explicit "laws" were pre-written. Once the simulation "runs," the initial variables and energies combine like infinite dice throws, shaping what we call physics.

One of the most provocative implications of MSP revolves around the notion of "glitches." In standard simulations created by coders, glitches arise from errors in pre-written rules, an oversight by the creator. By contrast, MSP suggests that our universe evolved from fundamental elements struggling into emergent laws rather than being governed by rigid, pre-scripted instructions. As a result, there is no top-down written code to produce errors, meaning that traditional "glitches" cannot exist. The process that created our universe was not linear code execution but an evolutionary melding of energies and seed variables, free from a grand design that could fail. This is why we cannot find a "glitch" in our universe: it was not programmed in a conventional sense, but rather took shape according to probabilistic self-organisation, circumventing the sort of mistakes that plague deliberately scripted simulations.

Among infinite possibilities, one particular run generated the laws and constants we observe, enabling, though not mandating, the evolution of biological life. Crucially, MSP holds that humans are merely an incidental product of this specific cosmic run, not its purpose or apex. While we labour under illusions of cosmic significance due to our evolutionary wiring, the deeper logic reveals that we are almost certainly inside a simulation operated by some advanced post-biological mind.

Moreover, we ourselves are on track to create further quantum universes, ensuring the cycle continues. Although these arguments suggest a high probability that we are simulated, absolute proof remains beyond our mind's capacity, as every aspect of our logic is constrained by the rules of our own cosmic "box." Thus, speculating on any reality beyond the simulation is futile.

Nevertheless, the theoretical probability of MSP being correct is so overwhelmingly high that we can, for all intents and purposes, regard our reality as an elaborate computational run, one among countless such worlds in the multiverse simulation paradigm.

MSP hinges on several foundational claims:

- **Everything is Mathematical**
  Every observed phenomenon, from subatomic particles to conscious thought, is wholly explainable by mathematical processes. Even consciousness, as humans label it, emerges from a combination of biology and algorithmic functions.

- **We Are Definitely in a Simulation**
  Because the technological arc of any sufficiently advanced civilization points toward the eventual creation of quantum universes, it follows that we ourselves are far more likely a product of such a simulation than a "true" baseline cosmos.

- **Laws Not Pre-Written but Emergent**
Our universe's constants, like the speed of light, Planck's constant, and gravitational strength, arose from initial "fundamental variables," akin to rolling an unbounded set of dice. The specific "roll" that led to atoms, galaxies, and life forms in our run could easily have produced different outcomes in other runs.

- **Humanity as One By-product**
Humans and life are not the core reason for this simulation. We are simply one of many emergent products. MSP rejects the notion of anthropocentrism, instead placing humans as peripheral in the cosmic tapestry.

- **No Extinction, Post-Human Creation**
Contrary to "doomsday" views, MSP claims human lineage will not vanish. Instead, as technology races ahead, a post-human era will spawn beings free of biology's physical chains. We will create quantum universes ourselves, continuing the cycle.

- **Cognitive Limits**
Because our mind evolved within these emergent laws, it cannot visualize or logically parse the "outside." Our entire reasoning is bounded by cosmic constraints. We cannot meaningfully step beyond the box of our simulation.

Deductive Reasoning Behind MSP

- **Premise 1: In any Universe, Technological Development Doesn't Cease**
Observing human history, fire to agriculture to digital computing, supports the idea that sentient species do not freeze at a given level. They push boundaries until near-catastrophe or transcendence.

- **Premise 2: Quantum Computing Allows Building Entire Universes**
Once quantum computers reach vast qubit capacities, modelling everything from chemical reactions to emergent neural nets becomes feasible. MSP suggests that eventually one can simulate not just molecules but entire cosmic evolutions.

- **Premise 3: Humans are Already on That Path**
Our technological progress is exponential, indicating we will almost certainly attempt universal-scale simulation. Thus, it stands to reason other advanced civilizations have done so, repeatedly.
- **Premise 4: If Universal Simulations Are Achievable, We Are Likely in One**
The final link is that if universal simulations are a repeated outcome, any random "conscious viewpoint" is probably from within a simulation rather than from any baseline "original."

Given the unstoppable push toward building entire simulated universes, MSP deduces that we are already participants in one. We do not occupy the original platform.

To quantify the likelihood that we inhabit such a simulation, define:

- p_{tech}: Probability that an advanced species (including us) will succeed in making cosmic-scale quantum simulations.
- N **:** The average number of full-universe simulations each advanced civilization runs.

The chance that we find ourselves in one of these many simulations, p_{tech}, can be sketched as:

$$p_{sim} = p_{tech} \times (1 - e^{-N})$$

$p_{tech} \approx \mathbf{1}$ if we believe that technology never hits a final barrier (MSP's stance).

If N is large (say, thousands or millions of simulations per civilization), then $(1 - e^{-N})$ nears 1.

Hence, $p_{sim} \approx p_{tech}$, Because MSP firmly believes humans do not go extinct, but rather become post-human, we assume near-certain mastery of cosmic-scale simulation. Therefore $p_{tech} \approx \mathbf{1}$.

Multiplying yields:

$$p_{sim} \approx 1 \times 1 = 1$$

Under MSP assumptions,

- Unstoppable tech progress and
- Repeated re-creation of universes

There is essentially a **100%** chance that our own consciousness belongs to a sophisticated simulation. Put differently, the probability that we are not in a simulation is effectively zero when measured against the inevitability of cosmic-scale quantum computing.

Conclusion

In the grand tapestry of existence, we have spent millennia convincing ourselves that there must be some transcendent meaning behind it all. Every culture, every era, has offered its parables, gods, or scientific frameworks to validate our sense of importance. Yet the deeper we dig into the bedrock of our reality, through equations, experiments, philosophical inquiry, the more we confront an unsettling possibility: it's all a construct. Our cosmos, for all its breath-taking scope, may amount to no more than a computational whim, a simulation hatched by forces or entities beyond our capacity to fathom. We want to believe that each sunrise and sunset, each heartbreak and triumph, is woven into a grand narrative. But perhaps we simply cling to illusions coded into the system, illusions that will vanish the moment our simulation window closes.

At first, this recognition can feel like a gut punch to everything we cherish. Why strive, why create art, why love deeply, if our every breath is but the side-effect of lines of code? Yet it is precisely this recognition that can free us from the weight of cosmic expectations. If our entire reality is a temporary projection, an ephemeral swirl of bits entangled in space-time, the notion of *meaning* becomes our own to define, or to disregard entirely. What a liberating prospect: no final judgment, no cosmic scoreboard, no supreme being scribbling down our transgressions. If it all dissolves into data once we exit the stage, then the joke is on anyone who wastes these fleeting hours chasing borrowed dreams.

Some will recoil at such bleakness, but maybe bleakness is just another coded sensation, another parameter in the grand simulation. If absolutely everything, fear, hope, longing, exaltation, is a bundle of qubits processed in some cosmic mainframe, then *why not* enjoy every moment to the fullest? Imagine liberating ourselves from the treadmill of presumed destiny, from the obsessive search for cosmic purpose. Your pains, your regrets, your ambitions, none of these are carved into the essence of the universe, because there **is** no essence. The single, inviolable truth is your immediate experience, *right now*. When you laugh, you create the only meaning that is real to you. When you ache with sorrow, that ache is the entire cosmos to your simulated mind. There is no hidden script, no preordained verdict. If the system's fundamental software tells us we can feel joy, then joy is as valid as any grand design.

In a world unshackled by deeper purpose, our so-called "problems" become optional. No cosmic ledger tallies our virtues. We can refuse the illusions of "achievement" and "legacy," illusions that matter only if we assume something outlasts us. They don't, because after we close our eyes for good, the universe does not owe us a postscript. If everything, from the swirl of galaxies to the nuance in your lover's voice, is ultimately a cosmic charade conjured in code, then the highest freedom might be to stop searching for reasons. Let the illusions serve you, instead of you serving them. Laugh because it's fun. Love because it enriches your fleeting sense of wonder. Cherish the ephemeral connections we form with others who share in this strange, pixelated dream.

And what of death? Ah, the question that has haunted philosophers since humans first woke to self-awareness. If we live in a simulation, death simply switches off your slot in the processing array. The cosmic CPU moves on, indifferent. No afterlife beckons, no final boss awaits. Perhaps your memories vanish with the collapse of your personal wave function. No ethereal soul floats upward to a paradise in the clouds. The simulation neither requires nor forbids such illusions; it simply does not care. You are free to conjure them if they soothe you, but know that these beliefs hold no more special cosmic currency than a child's bedtime story.

Paradoxically, it is precisely *because* nothing matters in any cosmic sense that everything can matter deeply on a personal level. Your tears at a funeral, your laughter at a friend's wedding, these are your truths, brief beacons of authenticity in a digital wilderness. If no grand scoreboard will immortalize your deeds, all the more reason to *savour every possible delight* in the short, hallucinatory window you call a lifetime. Sip your morning coffee like it's the last time it will ever pass your lips. Indulge in the small sensory miracles, music resonating in your ears, the warmth of the sun on your face, the electric pulse of a shared smile with someone who matters. Each moment is as much the "purpose" of the entire cosmos as anything else could be, precisely because none of it carries any inherent cosmic weight.

Of course, the simulation might have coded in a thousand illusions to keep us engaged, fear of meaninglessness, thirst for achievement, moral duty, love, hatred, artistic mania. But illusions can still be delicious, heartbreakingly real to the ones experiencing them. Why not treat them as ephemeral gifts, like fireworks that blaze brilliantly before vanishing into the night? Your life is that firework: sudden, sparkling, and *gone in an instant*. However unnerving it may sound, this ephemeral nature is precisely what makes it so exhilarating. The system doesn't supply you with an ultimate reason to exist; it merely provides the canvas. You get to paint, scribble, or set it ablaze at your whim.

Thus we conclude with a scandalous, liberating manifesto: We do not matter, and that *matters* so little that it grants us the highest freedom. If all that grounds us is a coded rule set, mortal illusions thrown upon the cosmic screen, then *why not* drink the nectar of existence to its fullest? Revel in love, in heartbreak, in ecstasy, in boredom, without ever demanding that it align with some cosmic script. You are the cosmic script: your laughter, your tears, your quiet walks under a starry sky that may be no more than pixelated illusions. Don't let the phantom chase for higher meaning shackle your short, electric spin through simulated existence.

Embrace the possibility that there is no "beyond death," that this life, these ephemeral tastes, touches, laughs, is all you ever get. Under that terrifying clarity, the choice is obvious: celebrate every sunrise, love recklessly, make memories with the ephemeral friends who share this cosmic charade. Resist the empty illusions of cosmic significance, and instead author your own fleeting significance. In this final understanding, you discover a scandalous kind of peace: *Nothing matters, so you are free to truly live.* Let that be your defiant anthem in a world that is, quite possibly, a grand and beautiful simulation.

www.ingramcontent.com/pod-product-compliance
Lightning Source LLC
LaVergne TN
LVHW090405210926
848480LV00002B/464